Use R!

Series Editors:
Robert Gentleman   Kurt Hornik   Giovanni G. Parmigiani

For further volumes:
http://www.springer.com/series/6991

Graham Williams

Data Mining with Rattle and R

The Art of Excavating Data for Knowledge Discovery

 Springer

Graham Williams
Togaware Pty Ltd
PO Box 655
Jamison Centre
ACT, 2614
Australia
Graham.Williams@togaware.com

Series Editors:
Robert Gentleman
Program in Computational Biology
Division of Public Health Sciences
Fred Hutchinson Cancer Research Center
1100 Fairview Avenue, N. M2-B876
Seattle, Washington 98109
USA

Kurt Hornik
Department of Statistik and Mathematik
Wirtschaftsuniversität Wien
Augasse 2-6
A-1090 Wien
Austria

Giovanni G. Parmigiani
The Sidney Kimmel Comprehensive
Cancer Center at Johns Hopkins University
550 North Broadway
Baltimore, MD 21205-2011
USA

ISBN 978-1-4419-9889-7   e-ISBN 978-1-4419-9890-3
DOI 10.1007/978-1-4419-9890-3
Springer New York Dordrecht Heidelberg London

Library of Congress Control Number: 2011934490

Springer is part of Springer Science+Business Media (www.springer.com)

To Catharina

Preface

Knowledge leads to wisdom and better understanding. Data mining builds knowledge from information, adding value to the ever-increasing stores of electronic data that abound today. Emerging from the database community in the late 1980s' data mining grew quickly to encompass researchers and technologies from machine learning, high-performance computing, visualisation, and statistics, recognising the growing opportunity to add value to data. Today, this multidisciplinary and transdisciplinary effort continues to deliver new techniques and tools for the analysis of very large collections of data. Working on databases that are now measured in the terabytes and petabytes, data mining delivers discoveries that can improve the way an organisation does business. Data mining enables companies to remain competitive in this modern, data-rich, information-poor, knowledge-hungry, and wisdom-scarce world. Data mining delivers knowledge to drive the getting of wisdom.

A wide range of techniques and algorithms are used in data mining. In performing data mining, many decisions need to be made regarding the choice of methodology, data, tools, and algorithms.

Throughout this book, we will be introduced to the basic concepts and algorithms of data mining. We use the free and open source software Rattle (Williams, 2009), built on top of the R statistical software package (R Development Core Team, 2011). As free software the source code of Rattle and R is available to everyone, without limitation. Everyone is permitted, and indeed encouraged, to read the source code to learn, understand verify, and extend it. R is supported by a worldwide network of some of the world's leading statisticians and implements all of the key algorithms for data mining.

This book will guide the reader through the various options that Rattle provides and serves to guide the new data miner through the use of Rattle. Many excursions into using R itself are presented, with the aim of encouraging readers to use R directly as a scripting language. Through scripting comes the necessary integrity and repeatability required for professional data mining.

Features

A key feature of this book, which differentiates it from many other very good textbooks on data mining, is the focus on the hands-on end-to-end process for data mining. We cover data understanding, data preparation, model building, model evaluation, data refinement, and practical deployment. Most data mining textbooks have their primary focus on just the model building—that is, the algorithms for data mining. This book, on the other hand, shares the focus with data and with model evaluation and deployment.

In addition to presenting descriptions of approaches and techniques for data mining using modern tools, we provide a very practical resource with actual examples using Rattle. Rattle is easy to use and is built on top of R. As mentioned above, we also provide excursions into the command line, giving numerous examples of direct interaction with R. The reader will learn to rapidly deliver a data mining project using software obtained for free from the Internet. Rattle and R deliver a very sophisticated data mining environment.

This book encourages the concept of programming with data, and this theme relies on some familiarity with the programming of computers. However, students without that background will still benefit from the material by staying with the Rattle application. All readers are encouraged, though, to consider becoming familiar with some level of writing commands to process and analyse data.

The book is accessible to many readers and not necessarily just those with strong backgrounds in computer science or statistics. At times, we do introduce more sophisticated statistical, mathematical, and computer science notation, but generally aim to keep it simple. Sometimes this means oversimplifying concepts, but only where it does not lose the intent of the concept and only where it retains its fundamental accuracy.

At other times, the presentation will leave the more statistically sophisticated wanting. As important as the material is, it is not always easily covered within the confines of a short book. Other resources cover such material in more detail. The reader is directed to the extensive mathematical treatment by Hastie et al. (2009). For a more introductory treatment using R for statistics, see Dalgaard (2008). For a broader perspective on using R, including a brief introduction to the tools in R for data mining, Adler (2010) is recommended. For an introduction to data mining with a case study orientation, see Torgo (2010).

Organisation

Chapter 1 sets the context for our data mining. It presents an overview of data mining, the process of data mining, and issues associated with data mining. It also canvasses open source software for data mining.

Chapter 2 then introduces Rattle as a graphical user interface (GUI) developed to simplify data mining projects. This covers the basics of interacting with R and Rattle, providing a quick-start guide to data mining.

Chapters 3 to 7 deal with data—we discuss the data, exploratory, and transformational steps of the data mining process. We introduce data and how to select variables and the partitioning of our data in Chapter 3. Chapter 4 covers the loading of data into Rattle and R. Chapters 5 and 6 then review various approaches to exploring the data in order for us to gain our initial insights about the data. We also learn about the distribution of the data and how to assess the appropriateness of any analysis. Often, our exploration of the data will lead us to identify various issues with the data. We thus begin cleaning the data, dealing with missing data, transforming the data, and reducing the data, as we describe in Chapter 7.

Chapters 8 to 14 then cover the building of models. This is the next step in data mining, where we begin to represent the knowledge discovered. The concepts of modelling are introduced in Chapter 8, introducing descriptive and predictive data mining. Specific descriptive data mining approaches are then covered in Chapters 9 (clusters) and 10 (association rules). Predictive data mining approaches are covered in Chapters 11 (decision trees), 12 (random forests), 13 (boosting), and 14 (support vector machines). Not all predictive data mining approaches are included, leaving some of the well-covered topics (including linear regression and neural networks) to other books.

Having built a model, we need to consider how to evaluate its performance. This is the topic for Chapter 15. We then consider the task of deploying our models in Chapter 16.

Appendix A can be consulted for installing R and Rattle. Both R and Rattle are open source software and both are freely available on multiple platforms. Appendix B describes in detail how the datasets used throughout the book were obtained from their sources and how they were transformed into the datasets made available through **rattle**.

Production and Typographical Conventions

This book has been typeset by the author using LaTeX and R's Sweave(). All R code segments included in the book are run at the time of typesetting the book, and the results displayed are directly and automatically obtained from R itself. The Rattle screen shots are also automatically generated as the book is typeset.

Because all R code and screen shots are automatically generated, the output we see in the book should be reproducible by the reader. All code is run on a 64 bit deployment of R on a Ubuntu GNU/Linux system. Running the same code on other systems (particularly on 32 bit systems) may result in slight variations in the results of the numeric calculations performed by R.

Other minor differences will occur with regard to the widths of lines and rounding of numbers. The following options are set when typesetting the book. We can see that `width=` is set to 58 to limit the line width for publication. The two options `scipen=` and `digits=` affect how numbers are presented:

```
> options(width=58, scipen=5, digits=4, continue="  ")
```

Sample code used to illustrate the interactive sessions using R will include the R prompt, which by default is "> ". However, we generally do not include the usual continuation prompt, which by default consists of "+ ". The continuation prompt is used by R when a single command extends over multiple lines to indicate that R is still waiting for input from the user. For our purposes, including the continuation prompt makes it more difficult to cut-and-paste from the examples in the electronic version of the book. The `options()` example above includes this change to the continuation prompt.

R code examples will appear as code blocks like the following example (though the continuation prompt, which is shown in the following example, will not be included in the code blocks in the book).

```
> library(rattle)

Rattle: A free graphical interface for data mining with R.
Version 2.6.7 Copyright (c) 2006-2011 Togaware Pty Ltd.
Type 'rattle()' to shake, rattle, and roll your data.

> rattle()

Rattle timestamp: 2011-06-13 09:57:52

> cat("Welcome to Rattle",
+     "and the world of Data Mining.\n")

Welcome to Rattle and the world of Data Mining.
```

In providing example output from commands, at times we will truncate the listing and indicate missing components with [...]. While most examples will illustrate the output exactly as it appears in R, there will be times where the format will be modified slightly to fit publication limitations. This might involve silently removing or adding blank lines.

In describing the functionality of Rattle, we will use a sans serif font to identify a Rattle widget (a graphical user interface component that we interact with, such as a button or menu). The kinds of widgets that are used in Rattle include the check box for turning options on and off, the radio button for selecting an option from a list of alternatives, file selectors for identifying files to load data from or to save data to, combo boxes for making selections, buttons to click for further plots or information, spin buttons for setting numeric options, and the text view, where the output from R commands will be displayed.

R provides very many *packages* that together deliver an extensive toolkit for data mining. **rattle** is itself an R package—we use a bold font to refer to R packages. When we discuss the functions or commands that we can type at the R prompt, we will include parentheses with the function name so that it is clearly a reference to an R function. The command `rattle()`, for example, will start the user interface for Rattle. Many functions and commands can also take arguments, which we indicate by trailing the argument with an equals sign. The `rattle()` command, for example, can accept the command argument `csvfile=`.

Implementing Rattle

Rattle has been developed using the Gnome (1997) toolkit with the Glade (1998) graphical user interface (GUI) builder. Gnome is independent of any programming language, and the GUI side of Rattle started out using the Python (1989) programming language. I soon moved to R directly, once **RGtk2** (Lawrence and Temple Lang, 2010) became available, providing access to Gnome from R. Moving to R allowed us to avoid the idiosyncrasies of interfacing multiple languages.

The Glade graphical interface builder is used to generate an XML file that describes the interface independent of the programming language. That file can be loaded into any supported programming language to display the GUI. The actual functionality underlying the application is then written in any supported language, which includes Java, C, C++, Ada, Python, Ruby, and R! Through the use of Glade, we have the freedom to quickly change languages if the need arises.

R itself is written in the procedural programming language C. Where computation requirements are significant, R code is often translated into C code, which will generally execute faster. The details are not important for us here, but this allows R to be surprisingly fast when it needs to be, without the users of R actually needing to be aware of how the function they are using is implemented.

Currency

New versions of R are released twice a year, in April and October. R is free, so a sensible approach is to upgrade whenever we can. This will ensure that we keep up with bug fixes and new developments, and we won't annoy the developers with questions about problems that have already been fixed.

The examples included in this book are from version 2.13.0 of R and version 2.6.7 of Rattle. Rattle is an ever-evolving package and, over time, whilst the concepts remain, the details will change. For example, the advent of **ggplot2** (Wickham, 2009) provides an opportunity to significantly develop its graphics capabilities. Similarly, **caret** (Kuhn et al., 2011) offers a newer approach to interfacing various data mining algorithms, and we may see Rattle take advantage of this. New data mining algorithms continue to emerge and may be incorporated over time.

Similarly, the screen shots included in this book are current only for the version of **Rattle** available at the time the book was typeset. Expect some minor changes in various windows and text views, and the occasional major change with the addition of new functionality.

Appendix A includes links to guides for installing **Rattle**. We also list there the versions of the primary packages used by **Rattle**, at least as of the date of typesetting this book.

Acknowledgements

This book has grown from a desire to share experiences in using and deploying data mining tools and techniques. A considerable proportion of the material draws on over 20 years of teaching data mining to undergraduate and graduate students and running industry-based courses. The aim is to provide recipe-type material that can be easily understood and deployed, as well as reference material covering the concepts and terminology a data miner is likely to come across.

Many thanks are due to students from the Australian National University, the University of Canberra, and elsewhere who over the years have been the reason for me to collect my thoughts and experiences with data mining and to bring them together into this book. I have benefited from their insights into how they learn best. They have also contributed in a number of ways with suggestions and example applications. I am also in debt to my colleagues over the years, particularly Peter Milne, Joshua Huang, Warwick Graco, John Maindonald, and Stuart Hamilton, for their support and contributions to the development of data mining in Australia.

Colleagues in various organisations deploying or developing skills in data mining have also provided significant feedback, as well as the motivation, for this book. Anthony Nolan deserves special mention for his enthusiasm and ongoing contribution of ideas that have helped fine-tune the material in the book.

Many others have also provided insights and comments. Illustrative examples of using R have also come from the R mailing lists, and I have used many of these to guide the kinds of examples that are included in the book. The many contributors to those lists need to be thanked.

Thanks also go to the reviewers, who have added greatly to the readability and usability of the book. These include Robert Muenchen, Peter Christen, Peter Helmsted, Bruce McCullough, and Balázs Bárány. Thanks also to John Garden for his encouragement and insights in choosing a title for the volume.

My very special thanks to my wife, Catharina, and children, Sean and Anita, who have endured my indulgence in bringing this book together.

Canberra   *Graham J. Williams*

Contents

Part I

Explorations

Chapter 1

Introduction

For the keen data miner, Chapter 2 provides a quick-start guide to data mining with Rattle, working through a sample process of loading a dataset and building a model.

Data mining is the art and science of intelligent data analysis. The aim is to discover meaningful insights and knowledge from **data**. Discoveries are often expressed as **models**, and we often describe data mining as the process of building models. A model captures, in some formulation, the essence of the discovered knowledge. A model can be used to assist in our **understanding** of the world. Models can also be used to make **predictions**.

For the data miner, the discovery of new knowledge and the building of models that nicely predict the future can be quite rewarding. Indeed, data mining should be exciting and fun as we watch new insights and knowledge emerge from our data. With growing enthusiasm, we meander through our data analyses, following our intuitions and making new discoveries all the time—discoveries that will continue to help change our world for the better.

Data mining has been applied in most areas of endeavour. There are data mining teams working in business, government, financial services, biology, medicine, risk and intelligence, science, and engineering. Anywhere we collect data, data mining is being applied and feeding new knowledge into human endeavour.

We are living in a time where data is collected and stored in unprecedented volumes. Large and small government agencies, commercial enterprises, and noncommercial organisations collect data about their businesses, customers, human resources, products, manufacturing processes, suppliers, business partners, local and international markets, and competitors. Data is the fuel that we inject into the data mining engine.

Turning data into information and then turning that information into knowledge remains a key factor for "success." Data contains valuable information that can support managers in their business decisions to effectively and efficiently run a business. Amongst data there can be hidden clues of the fraudulent activity of criminals. Data provides the basis for understanding the scientific processes that we observe in our world. Turning data into information is the basis for identifying new opportunities that lead to the discovery of new knowledge, which is the linchpin of our society!

Data mining is about building models from data. We build models to gain insights into the world and how the world works so we can predict how things behave. A data miner, in building models, deploys many different data analysis and model building techniques. Our choices depend on the business problems to be solved. Although data mining is not the only approach, it is becoming very widely used because it is well suited to the data environments we find in today's enterprises. This is characterised by the volume of data available, commonly in the gigabytes and terabytes and fast approaching the petabytes. It is also characterised by the complexity of that data, both in terms of the relationships that are awaiting discovery in the data and the data types available today, including text, image, audio, and video. The business environments are also rapidly changing, and analyses need to be performed regularly and models updated regularly to keep up with today's dynamic world.

Modelling is what people often think of when they think of data mining. Modelling is the process of turning data into some structured form or model that reflects the supplied data in some useful way. Overall, the aim is to explore our data, often to address a specific problem, by modelling the world. From the models, we gain new insights and develop a better understanding of the world.

Data mining, in reality, is so much more than simply modelling. It is also about understanding the business context within which we deploy it. It is about understanding and collecting data from across an enterprise and from external sources. It is then about building models and evaluating them. And, most importantly, it is about deploying those models to deliver benefits.

There is a bewildering array of tools and techniques at the disposal of the data miner for gaining insights into data and for building models.

This book introduces some of these as a starting point on a longer journey to becoming a practising data miner.

1.1 Data Mining Beginnings

Data mining, as a named endeavour, emerged at the end of the 1980s from the database community, which was wondering where the next big steps forward were going to come from. Relational database theory had been developed and successfully deployed, and thus began the era of collecting large amounts of data. How do we add value to our massive stores of data?

The first few data mining workshops in the early 1990s attracted the database community researchers. Before long, other computer science, and particularly artificial intelligence, researchers began to get interested. It is useful to note that a key element of "intelligence" is the ability to learn, and machine learning research had been developing technology for this for many years. Machine learning is about collecting observational data through interacting with the world and building models of the world from such data. That is pretty much what data mining was also setting about to do. So, naturally, the machine learning and data mining communities started to come together.

However, statistics is one of the fundamental tools for data analysis, and has been so for over a hundred years. Statistics brings to the table essential ideas about uncertainty and how to make allowances for it in the models that we build. Statistics provides a framework for understanding the "strength" or veracity of models that we might build from data. Discoveries need to be statistically sound and statistically significant, and any uncertainty associated with the modelling needs to be understood. Statistics plays a key role in today's data mining.

Today, data mining is a discipline that draws on sophisticated skills in computer science, machine learning, and statistics. However, a data miner will work in a team together with data and domain experts.

1.2 The Data Mining Team

Many data mining projects work with ill-defined and ambiguous goals. Whilst the first reaction to such an observation is that we should become better at defining the problem, the reality is that often the problem to be solved is identified and refined as the data mining project progresses. That's natural.

An initiation meeting of a data mining project will often involve data miners, *domain experts*, and *data experts*. The data miners bring the statistical and algorithmic understanding, programming skills, and key investigative ability that underlies any analysis. The domain experts know about the actual problem being tackled, and are often the business experts who have been working in the area for many years. The data experts know about the data, how it has been collected, where it has been stored, how to access and combine the data required for the analysis, and any idiosyncrasies and data traps that await the data miner.

Generally, neither the domain expert nor the data expert understand the needs of the data miner. In particular, as a data miner we will often find ourselves encouraging the data experts to provide (or to provide access to) all of the data, and not just the data the data expert thinks might be useful. As data miners we will often think of ourselves as "greedy" consumers of all the data we can get our hands on.

It is critical that all three experts come together to deliver a data mining project. Their different understandings of the problem to be tackled all need to meld to deliver a common pathway for the data mining project. In particular, the data miner needs to understand the problem domain perspective and understand what data is available that relates to the problem and how to get that data, and identify what data processing is required prior to modelling.

1.3   Agile Data Mining

Building models is only one of the tasks that the data miner performs. There are many other important tasks that we will find ourselves involved in. These include ensuring our data mining activities are tackling the right problem; understanding the data that is available, turning noisy data into data from which we can build robust models; evaluating and demonstrating the performance of our models; and ensuring the effective deployment of our models.

Whilst we can easily describe these steps, it is important to be aware that data mining is an agile activity. The concept of agility comes from the agile software engineering principles, which include the evolution or incremental development of the problem requirements, the requirement for regular client input or feedback, the testing of our models as they are being developed, and frequent rebuilding of the models to improve their performance.

An allied aspect is the concept of pair programming, where two data miners work together on the same data in a friendly, competitive, and collaborative approach to building models. The agile approach also emphasises the importance of face-to-face communication, above and beyond all of the effort that is otherwise often expended, and often wasted, on written documentation. This is not to remove the need to write documents but to identify what is really required to be documented.

We now identify the common steps in a data mining project and note that the following chapters of this book then walk us through these steps one step at a time!

1.4   The Data Mining Process

The Cross Industry Process for Data Mining (CRISP-DM, 1996) provides a common and well-developed framework for delivering data mining projects. CRISP-DM identifies six steps within a typical data mining project:

1. Problem Understanding

2. Data Understanding

3. Data Preparation

4. Modeling

5. Evaluation

6. Deployment

The chapters in this book essentially follow this step-by-step process of a data mining project, and Rattle is very much based around these same steps. Using a tab-based interface, each tab represents one of the steps, and we proceed through the tabs as we work our way through a data mining project. One noticeable exception to this is the first step, problem understanding. That is something that needs study, discussion, thought, and brain power. Practical tools to help in this process are not common.

1.5   A Typical Journey

Many organisations are looking to set up a data mining capability, often called the analytics team. Within the organisation, data mining projects can be initiated by the business or by this analytics team. Often, for best business engagement, a business-initiated project works best, though business is not always equipped to understand where data mining can be applied. It is often a mutual journey.

Data miners, by themselves, rarely have the deeper knowledge of business that a professional from the business itself has. Yet the business owner will often have very little knowledge of what data mining is about, and indeed, given the hype, may well have the wrong idea. It is not until they start getting to see some actual data mining models for their business that they start to understand the project, the possibilities, and a glimpse of the potential outcomes.

We will relate an actual experience over six months with six significant meetings of the business team and the analytics team. The picture we paint here is a little simplified and idealised but is not too far from reality.

Meeting One The data miners sit in the corner to listen and learn. The business team understands little about what the data miners might be able to deliver. They discuss their current business issues and steps being taken to improve processes. The data miners have little to offer just yet but are on the lookout for the availability of data from which they can learn.

Meeting Two The data miners will now often present some observations of the data from their initial analyses. Whilst the analyses might be well presented graphically, and are perhaps interesting, they are yet to deliver any new insights into the business. At least the data miners are starting to get the idea of the business, as far as the business team is concerned.

Meeting Three The data miners start to demonstrate some initial modelling outcomes. The results begin to look interesting to the business team. They are becoming engaged, asking questions, and understanding that the data mining team has uncovered some interesting insights.

Meeting Four The data miners are the main agenda item. Their analyses are starting to ring true. They have made some quite interesting discoveries from the data that the business team (the domain and data experts) supplied. The discoveries are nonobvious, and sometimes intriguing. Sometimes they are also rather obvious.

Meeting Five The models are presented for evaluation. The data mining team has presented its evaluation of how well the models perform and explained the context for the deployment of the models. The business team is now keen to evaluate the model on real cases and monitor its performance over a period of time.

Meeting Six The models have been deployed into business and are being run daily to match customers and products for marketing, to identify insurance claims or credit card transactions that may be fraudulent, or taxpayers whose tax returns may require refinement. Procedures are in place to monitor the performance of the model over time and to sound alarm bells once the model begins to deviate from expectations.

The key to much of the data mining work described here, in addition to the significance of communication, is the reliance and focus on data. This leads us to identify some key principles for data mining.

1.6   Insights for Data Mining

The starting point with all data mining is the data. We need to have good data that relates to a process that we wish to understand and improve. Without data we are simply guessing.

Considerable time and effort spent getting our data into shape is a key factor in the success of a data mining project. In many circumstances, once we have the right data for mining, the rest is straightforward. As many others note, this effort in data collection and data preparation can in fact be the most substantial component of a data mining project.

My list of insights for data mining, in no particular order, includes:

1. Focus on the data and understand the business.

2. Use training/validate/test datasets to build/tune/evaluate models.

3. Build multiple models: most give very similar performance.

4. Question the "perfect" model as too good to be true.

5. Don't overlook how the model is to be deployed.

6. Stress repeatability and efficiency, using scripts for everything.

7. Let the data talk to you but not mislead you.

8. Communicate discoveries effectively and visually.

1.7   Documenting Data Mining

An important task whilst data mining is the recording of the process. We need to be vigilant to record all that is done. This is often best done through the code we write to perform the analysis rather than having to document the process separately. Having a separate process to document the data mining will often mean that it is rarely completed. An implication of this is that we often capture the process as transparent, executable code rather than as a list of instructions for using a GUI.

There are many important advantages to ensuring we document a project through our coding of the data analyses. There will be times when we need to hand a project to another data miner. Or we may cease work on a project for a period of time and return to it at a later stage. Or we have performed a series of analyses and much the same process will need to be repeated again in a year's time. For whatever reason, when we return to a project, we find the documentation, through the coding, essential in being efficient and effective data miners.

Various things should be documented, and most can be documented through a combination of code and comments. We need to document our access to the source data, how the data was transformed and cleaned, what new variables were constructed, and what summaries were generated to understand the data. Then we also need to record how we built models and what models were chosen and considered. Finally, we record the evaluation and how we collect the data to support the benefit that we propose to obtain from the model.

Through documentation, and ideally by developing documented code that tells the story of the data mining project and the actual process as well, we will be communicating to others how we can mine data. Our processes can be easily reviewed, improved, and automated. We can transparently stand behind the results of the data mining by having openly available the process and the data that have led to the results.

1.8   Tools for Data Mining: R

R is used throughout this book to illustrate data mining procedures. It is the programming language used to implement the Rattle graphical user interface for data mining. If you are moving to R from SAS or SPSS, then you will find Muenchen (2008) a great resource.[1]

R is a sophisticated statistical software package, easily installed, instructional, state-of-the-art, and it is free and open source. It provides all of the common, most of the less common, and all of the new approaches to data mining.

The basic modus operandi in using R is to write scripts using the R language. After a while you will want to do more than issue single simple commands and rather write programs and systems for common tasks that suit your own data mining. Thus, saving our commands to an R script file (often with the .R filename extension) is important. We can then rerun our scripts to transform our source data, at will and automatically, into information and knowledge. As we progress through the book, we will become familiar with the common R functions and commands that we might combine into a script.

Whilst for data mining purposes we will focus on the use of the Rattle GUI, more advanced users might prefer the powerful Emacs editor, augmented with the ESS package, to develop R code directly. Both run under GNU/Linux, Mac/OSX, and Microsoft Windows.

We also note that direct interaction with R has a steeper learning curve than using GUI based systems, but once over the hurdle, performing operations over the same or similar datasets becomes very easy using its programming language interface.

A paradigm that is encouraged throughout this book is that of *learning by example* or *programming by example* (Cypher, 1993). The intention is that anyone will be able to easily replicate the examples from the book and then fine-tune them to suit their own needs. This is one of the underlying principles of Rattle, where all of the R commands that are used under the graphical user interface are also exposed to the user. This makes it a useful teaching tool in learning R for the specific task of data mining, and also a good memory aid!

1.9 Tools for Data Mining: Rattle

Rattle is built on the statistical language R, but an understanding of R is not required in order to use it. Rattle is simple to use, quick to deploy, and allows us to rapidly work through the data processing, modelling, and evaluation phases of a data mining project. On the other hand,

---

[1] An early version is available from http://r4stats.com.

R provides a very powerful language for performing data mining well beyond the limitations that are embodied in any graphical user interface and the consequently canned approaches to data mining. When we need to fine-tune and further develop our data mining projects, we can migrate from Rattle to R.

Rattle can save the current state of a data mining task as a Rattle project. A Rattle project can then be loaded at a later time or shared with other users. Projects can be loaded, modified, and saved, allowing check pointing and parallel explorations. Projects also retain all of the R code for transparency and repeatability. This is an important aspect of any scientific and deployed endeavour—to be able to repeat our "experiments."

Whilst a user of Rattle need not necessarily learn R, Rattle exposes all of the underlying R code to allow it to be directly deployed within the R Console as well as saved in R scripts for future reference. The R code can be loaded into R (outside of Rattle) to repeat any data mining task.

Rattle by itself may be sufficient for all of a user's needs, particularly in the context of introducing data mining. However, it also provides a stepping stone to more sophisticated processing and modelling in R itself. It is worth emphasising that the user is not limited to how Rattle does things. For sophisticated and unconstrained data mining, the experienced user will progress to interacting directly with R.

The typical workflow for a data mining project was introduced above. In the context of Rattle, it can be summarised as:

1. Load a **Data**set.

2. **Select** variables and entities for exploring and mining.

3. **Explore** the data to understand how it is distributed or spread.

4. **Transform** the data to suit our data mining purposes.

5. Build our **Model**s.

6. **Evaluate** the models on other datasets.

7. **Export** the models for deployment.

It is important to note that at any stage the next step could well be a step to a previous stage. Also, we can save the contents of Rattle's **Log** tab as a repeatable record of the data mining process.

We illustrate a typical workflow that is embodied in the Rattle interface in Figure 1.1.

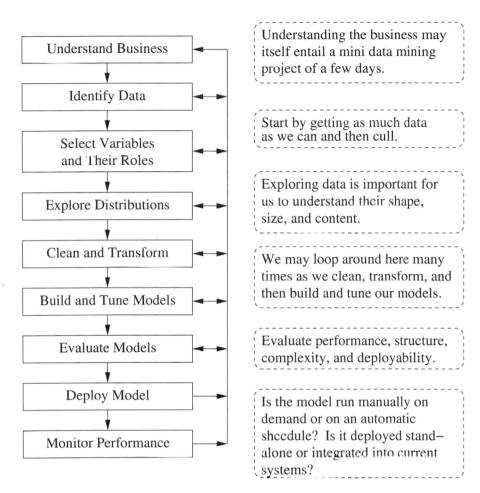

Figure 1.1: The typical workflow of a data mining project as supported by Rattle.

1.10   Why **R** and **Rattle**?

R and Rattle are free software in terms of allowing anyone the freedom to do as they wish with them. This is also referred to as open source software to distinguish it from closed source software, which does not provide the source code. Closed source software usually has quite restrictive licenses associated with it, aimed at limiting our freedom using it. This is separate from the issue of whether the software can be obtained for free (which is often, but not necessarily, the case for open source software) or must be purchased. R and Rattle can be obtained for free.

On 7 January 2009, the New York Times carried a front page technology article on R where a vendor representative was quoted:

> I think it addresses a niche market for high-end data analysts that want free, readily available code. ...We have customers who build engines for aircraft. I am happy they are not using freeware when I get on a jet.

This is a common misunderstanding about the concept of free and open source software. R, being free and open source software, is in fact a peer-reviewed software product that a number of the worlds top statisticians have developed and others have reviewed. In fact, anyone is permitted to review the R source code. Over the years, many bugs and issues have been identified and rectified by a large community of developers and users.

On the other hand, a closed source software product cannot be so readily and independently verified or viewed by others at will. Bugs and enhancement requests need to be reported back to the vendor. Customers then need to rely on a very select group of vendor-chosen people to assure the software, rectify any bugs in it, and enhance it with new algorithms. Bug fixes and enhancements can take months or years, and generally customers need to purchase the new versions of the software.

Both scenarios (open source and closed source) see a lot of effort put into the quality of their software. With open source, though, we all share it, whereas we can share and learn very little about the algorithms we use from closed source software.

It is worthwhile to highlight another reason for using R in the context of free and commercial software. In obtaining any software, due diligence is required in assessing what is available. However, what is finally delivered may be quite different from what was promised or even possible with the software, whether it is open source or closed source, free or commercial. With free open source software, we are free to use it without restriction. If we find that it does not serve our purposes, we can move on with minimal cost. With closed source commercial purchases, once the commitment is made to buy the software and it turns out not to meet our requirements, we are generally stuck with it, having made the financial commitment, and have to make do.

Moving back to R specifically, many have identified the pros and cons of using this statistical software package. We list some of the advantages with using R:

- *R is the most comprehensive statistical analysis package available.* It incorporates all of the standard statistical tests, models, and analyses, as well as providing a comprehensive language for managing and manipulating data. New technology and ideas often appear first in R.

- R is a programming language and environment developed for statistical analysis by practising statisticians and researchers. It reflects well on a very competent community of computational statisticians.

- R is now maintained by a core team of some 19 developers, including some very senior statisticians.

- The graphical capabilities of R are outstanding, providing a fully programmable graphics language that surpasses most other statistical and graphical packages.

- The validity of the R software is ensured through openly validated and comprehensive governance as documented for the US Food and Drug Administration (R Foundation for Statistical Computing, 2008). Because R is open source, unlike closed source software, it has been reviewed by many internationally renowned statisticians and computational scientists.

- R is free and open source software, allowing anyone to use and, importantly, to modify it. R is licensed under the GNU General Public License, with copyright held by The R Foundation for Statistical Computing.

- R has no license restrictions (other than ensuring our freedom to use it at our own discretion), and so we can run it anywhere and at any time, and even sell it under the conditions of the license.

- Anyone is welcome to provide bug fixes, code enhancements, and new packages, and the wealth of quality packages available for R is a testament to this approach to software development and sharing.

- R has over 4800 packages available from multiple repositories specialising in topics like econometrics, data mining, spatial analysis, and bio-informatics.

- R is cross-platform. R runs on many operating systems and different hardware. It is popularly used on GNU/Linux, Macintosh, and Microsoft Windows, running on both 32 and 64 bit processors.

- R plays well with many other tools, importing data, for example, from CSV files, SAS, and SPSS, or directly from Microsoft Excel, Microsoft Access, Oracle, MySQL, and SQLite. It can also produce graphics output in PDF, JPG, PNG, and SVG formats, and table output for LATEX and HTML.

- R has active user groups where questions can be asked and are often quickly responded to, often by the very people who developed the environment—this support is second to none. Have you ever tried getting support from the core developers of a commercial vendor?

- New books for R (the Springer Use R! series) are emerging, and there is now a very good library of books for using R.

Whilst the advantages might flow from the pen with a great deal of enthusiasm, it is useful to note some of the disadvantages or weaknesses of R, even if they are perhaps transitory!

- R has a steep learning curve—it does take a while to get used to the power of R—but no steeper than for other statistical languages.

- R is not so easy to use for the novice. There are several simple-to-use graphical user interfaces (GUIs) for R that encompass point-and-click interactions, but they generally do not have the polish of the commercial offerings.

- Documentation is sometimes patchy and terse, and impenetrable to the non-statistician. However, some very high-standard books are increasingly plugging the documentation gaps.

- The quality of some packages is less than perfect, although if a package is useful to many people, it will quickly evolve into a very robust product through collaborative efforts.

- There is, in general, no one to complain to if something doesn't work. R is a software application that many people freely devote their own time to developing. Problems are usually dealt with quickly on the open mailing lists, and bugs disappear with lightning speed. Users who do require it can purchase support from a number of vendors internationally.

- Many R commands give little thought to memory management, and so R can very quickly consume all available memory. This can be a restriction when doing data mining. There are various solutions, including using 64 bit operating systems that can access much more memory than 32 bit ones.

1.11 Privacy

Before closing out our introduction to data mining and tools for doing it, we need to touch upon the topic of privacy. Laws in many countries can directly affect data mining, and it is very worthwhile to be aware of them and their penalties, which can often be severe.

There are basic principles relating to the protection of privacy that we should adhere to. Some are captured by the privacy principles developed by the international Organisation for Economic Co-operation and Development—the OECD (Organisation for Economic Co-operation and Development (OECD), 1980). They include:

- **Collection limitation**
  Data should be obtained lawfully and fairly, while some very sensitive data should not be held at all.

- **Data quality**
  Data should be relevant to the stated purposes, accurate, complete, and up-to-date; proper precautions should be taken to ensure this accuracy.

- **Purpose specification**
  The purposes for which data will be used should be identified, and the data should be destroyed if it no longer serves the given purpose.

- **Use limitation**
  Use of data for purposes other than specified is forbidden.

As data miners, we have a social responsibility to protect our society and individuals for the good and benefit of all of society. Please take that responsibility seriously. Think often and carefully about what you are doing.

1.12 Resources

This book does not attempt to be a comprehensive introduction to using R. Some basic familiarity with R will be gained through our travels in data mining using the Rattle interface and some excursions into R. In this respect, most of what we need to know about R is contained within the book. But there is much more to learn about R and its associated packages. We do list and comment on here a number of books that provide an entrée to R.

A good starting point for handling data in R is *Data Manipulation with R* (Spector, 2008). The book covers the basic data structures, reading and writing data, subscripting, manipulating, aggregating, and reshaping data.

Introductory Statistics with R (Dalgaard, 2008), as mentioned earlier, is a good introduction to statistics using R. *Modern Applied Statistics with S* (Venables and Ripley, 2002) is quite an extensive introduction to statistics using R. Moving more towards areas related to data mining, *Data Analysis and Graphics Using R* (Maindonald and Braun, 2007) provides excellent practical coverage of many aspects of exploring and modelling data using R. *The Elements of Statistical Learning* (Hastie et al., 2009) is a more mathematical treatise, covering all of the machine learning techniques discussed in this book in quite some mathematical depth. If you are coming to R from a SAS or SPSS background, then *R for SAS and SPSS Users* (Muenchen, 2008) is an excellent choice. Even if you are not a SAS or SPSS user, the book provides a straightforward introduction to using R.

Quite a few specialist books using R are now available, including *Lattice: Multivariate Data Visualization with R* (Sarkar, 2008), which covers the extensive capabilities of one of the graphics/plotting packages available for R. A newer graphics framework is detailed in *ggplot2: Elegant Graphics for Data Analysis* (Wickham, 2009). Bivand et al. (2008) cover applied spatial data analysis, Kleiber and Zeileis (2008) cover applied econometrics, and Cowpertwait and Metcalfe (2009) cover time series, to name just a few books in the R library.

Moving on from R itself and into data mining, there are very many general introductions available. One that is commonly used for teaching in computer science is Han and Kamber (2006). It provides a comprehensive generic introduction to most of the algorithms used by a data miner. It is presented at a level suitable for information technology and database graduates.

Chapter 2

Getting Started

New ideas are often most effectively understood and appreciated by actually doing something with them. So it is with data mining. Fundamentally, data mining is about practical application—application of the algorithms developed by researchers in artificial intelligence, machine learning, computer science, and statistics. This chapter is about getting started with data mining.

Our aim throughout this book is to provide hands-on practise in data mining, and to do so we need some computer software. There is a choice of software packages available for data mining. These include commercial closed source software (which is also often quite expensive) as well as free open source software. Open source software (whether freely available or commercially available) is *always* the best option, as it offers us the freedom to do whatever we like with it, as discussed in Chapter 1. This includes extending it, verifying it, tuning it to suit our needs, and even selling it. Such software is often of higher quality than commercial closed source software because of its open nature.

For our purposes, we need some good tools that are freely available to everyone and can be freely modified and extended by anyone. Therefore we use the open source and free data mining tool Rattle, which is built on the open source and free statistical software environment R. See Appendix A for instructions on obtaining the software. Now is a good time to install R. Much of what follows for the rest of the book, and specifically this chapter, relies on interacting with R and Rattle.

We can, quite quickly, begin our first data mining project, with Rattle's support. The aim is to build a model that captures the essence of the knowledge discovered from our data. Be careful though—there is a lot of effort required in getting our data into shape. Once we have quality data, Rattle can build a model with just four mouse clicks, but the effort is in preparing the data and understanding and then fine-tuning the models.

In this chapter, we use Rattle to build our first data mining model—a simple decision tree model, which is one of the most common models in data mining. We cover starting up (and quitting from) R, an overview of how we interact with Rattle, and then how to load a dataset and build a model. Once the enthusiasm for building a model is satisfied, we then review the larger tasks of understanding the data and evaluating the model. Each element of Rattle's user interface is then reviewed before we finish by introducing some basic concepts related to interacting directly with and writing instructions for R.

2.1   Starting R

R is a command line tool that is initiated either by typing the letter R (capital R—R is case-sensitive) into a command line window (e.g., a terminal in GNU/Linux) or by opening R from the desktop icon (e.g., in Microsoft Windows and Mac/OSX). This assumes that we have already installed R, as detailed in Appendix A.

One way or another, we should see a window (Figure 2.1) displaying the R prompt (>), indicating that R is waiting for our commands. We will generally refer to this as the R Console.

The Microsoft Windows R Console provides additional menus specifically for working with R. These include options for working with script files, managing packages, and obtaining help.

We start Rattle by loading **rattle** into the R library using `library()`. We supply the name of the package to load as the argument to the command. The `rattle()` command is then entered with an empty argument list, as shown below. We will then see the Rattle GUI displayed, as in Figure 2.2.

```
> library(rattle)
> rattle()
```

The Rattle user interface is a simple tab-based interface, with the idea being to work from the leftmost tab to the rightmost tab, mimicking the typical data mining process.

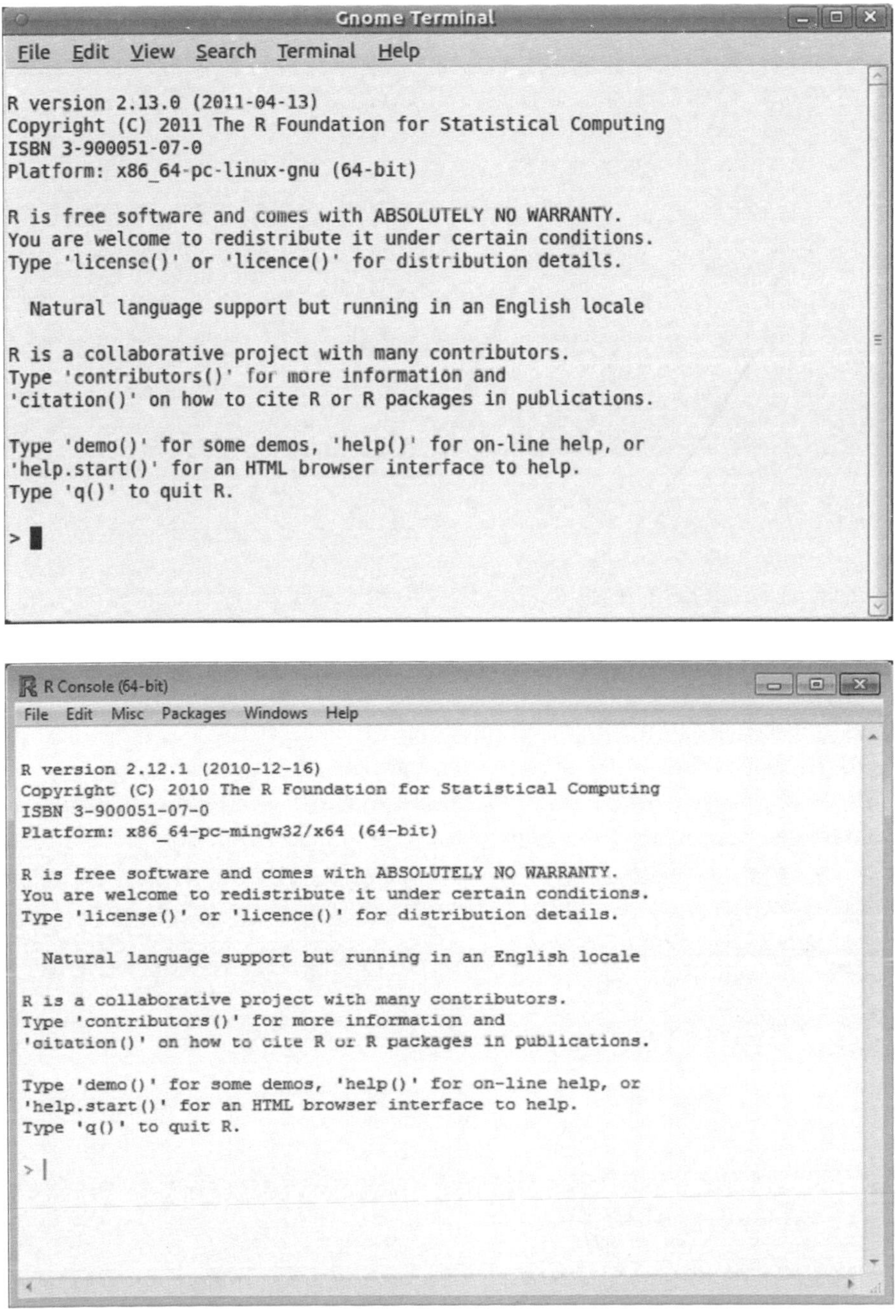

```
Gnome Terminal                                          _ □ ✕

File   Edit   View   Search   Terminal   Help

R version 2.13.0 (2011-04-13)
Copyright (C) 2011 The R Foundation for Statistical Computing
ISBN 3-900051-07-0
Platform: x86_64-pc-linux-gnu (64-bit)

R is free software and comes with ABSOLUTELY NO WARRANTY.
You are welcome to redistribute it under certain conditions.
Type 'license()' or 'licence()' for distribution details.

  Natural language support but running in an English locale

R is a collaborative project with many contributors.
Type 'contributors()' for more information and
'citation()' on how to cite R or R packages in publications.

Type 'demo()' for some demos, 'help()' for on-line help, or
'help.start()' for an HTML browser interface to help.
Type 'q()' to quit R.

> █
```

```
R  R Console (64-bit)                                      ─ □ ✕

File   Edit   Misc   Packages   Windows   Help

R version 2.12.1 (2010-12-16)
Copyright (C) 2010 The R Foundation for Statistical Computing
ISBN 3-900051-07-0
Platform: x86_64-pc-mingw32/x64 (64-bit)

R is free software and comes with ABSOLUTELY NO WARRANTY.
You are welcome to redistribute it under certain conditions.
Type 'license()' or 'licence()' for distribution details.

  Natural language support but running in an English locale

R is a collaborative project with many contributors.
Type 'contributors()' for more information and
'citation()' on how to cite R or R packages in publications.

Type 'demo()' for some demos, 'help()' for on-line help, or
'help.start()' for an HTML browser interface to help.
Type 'q()' to quit R.

> |
```

Figure 2.1: The R Console for GNU/Linux and Microsoft Windows. The prompt indicates that R is awaiting user commands.

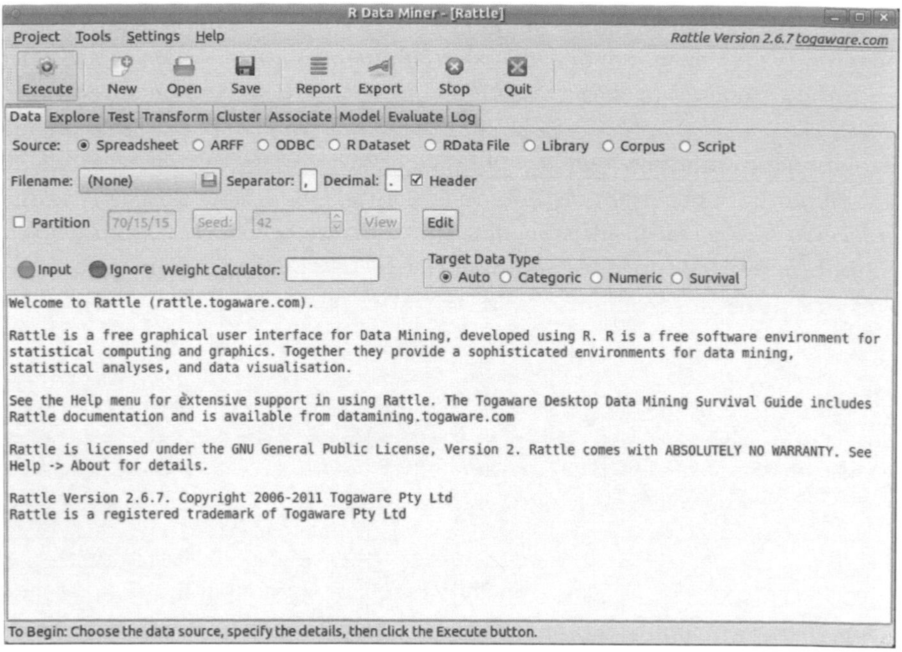

Figure 2.2: The initial Rattle window displays a welcome message and a little introduction to Rattle and R.

> **Tip:** *The key to using* **Rattle**, *as hinted at in the status bar on starting up* **Rattle**, *is to supply the appropriate information for a particular tab and to then* **click the Execute button** *to perform the action. Always make sure you have clicked the* **Execute** *button before proceeding to the next step.*

2.2 Quitting **Rattle** and **R**

A rather important piece of information, before we get into the details, is how to quit from the applications. To exit from Rattle, we simply click the Quit button. In general, this won't terminate the R Console. For R, the startup message (Figure 2.1) tells us to type q() to quit. We type this command into the R Console, including the parentheses so that the command is invoked rather than simply listing its definition. Pressing Enter will then ask R to quit:

```
> q()
Save workspace image? [y/n/c]:
```

We are prompted to save our workspace image. The workspace refers to all of the datasets and any other objects we have created in the current R session. We can save all of the objects currently available in a workspace between different invocations of R. We do so by choosing the y option. We might be in the middle of some complex analysis and wish to resume it at a later time, so this option is useful.

Many users generally answer n each time here, having already captured their analyses into script files. Script files allow us to automatically regenerate the results as required, and perhaps avoid saving and managing very large workspace files.

If we do not actually want to quit, we can answer c to cancel the operation and return to the R Console.

2.3   First Contact

In Chapter 1, we identified that a significant amount of effort within a data mining project is spent in processing our data into a form suitable for data mining. The amount of such effort should not be underestimated, but we do skip this step for now.

Once we have processed our data, we are ready to build a model—and with Rattle we can build the model with just a few mouse clicks. Using a sample dataset that someone else has already prepared for us, in Rattle we simply:

1. Click on the Execute button.
   Rattle will notice that no dataset has been identified, so it will take action, as in the next step, to ensure we have some data. This is covered in detail in Section 2.4 and Chapter 4.

2. Click on Yes within the resulting popup.
   The *weather* dataset is provided with Rattle as a small and simple dataset to explore the concepts of data mining. The dataset is described in detail in Chapter 3.

3. Click on the Model tab.
   This will change the contents of Rattle's main window to display options and information related to the building of models. This is where we tell Rattle what kind of model we want to build and how it should be built. The Model tab is described in more detail in Section 2.5, and model building is discussed in considerable detail in Chapters 8 to 14.

4. Click on the Execute button.

   Once we have specified what we want done, we ask Rattle to do it
   by clicking the Execute button. For simple model builders for small
   datasets, Rattle will only take a second or two before we see the
   results displayed in the text view window.

The resulting decision tree model, displayed textually in Rattle's text
view, is based on a sample dataset of historic daily weather observations
(the curious can skip a few pages ahead to see the actual decision tree in
Figure 2.5 on page 30).

   The data comes from a weather monitoring station located in Canberra, Australia, via the Australian Bureau of Meteorology. Each observation is a summary of the weather conditions on a particular day. It
has been processed to include a target variable that indicates whether it
rained the day following the particular observation. Using this historic
data, we have built a model to predict whether it will rain tomorrow.
Weather data is commonly available, and you might be able to build a
similar model based on data from your own region.

   With only one or two more clicks, further models can be built. A few
more clicks and we have an evaluation chart displaying the performance
of the model. Then, with just a click or two more, we will have the model
applied to a new dataset to generate scores for new observations.

   Now to the details. We will continue to use Rattle and also the simple
command line facility. The command line is not strictly necessary in
using Rattle, but as we develop our data mining capability, it will become
useful. We will load data into Rattle and explain the model that we have
built. We will build a second model and compare their performances.
We will then apply the model to a new dataset to provide scores for
a collection of new observations (i.e., predictions of the likelihood of it
raining tomorrow).

2.4   Loading a Dataset

With Rattle we can load a sample dataset in preparation for modelling, as
we have just done. Now we want to illustrate loading any data (perhaps
our own data) into Rattle.

   If we have followed the four steps in Section 2.3, then we will now
need to reset Rattle. Simply click the New button within the toolbar.
We are asked to confirm that we would like to clear the current project.

Alternatively, we might have exited Rattle and R, as described in Section 2.1, and need to restart everything, as also described in Section 2.1. Either way, we need to have a fresh Rattle ready so that we can follow the examples below.

On starting Rattle, we can, without any other action, click the Execute button in the toolbar. Rattle will notice that no CSV file (the default data format) has been specified (notice the "(None)" in the Filename: chooser) and will ask whether we wish to use one of the sample datasets supplied with the package. Click on Yes to do so, to see the data listed, as shown in Figure 2.3.

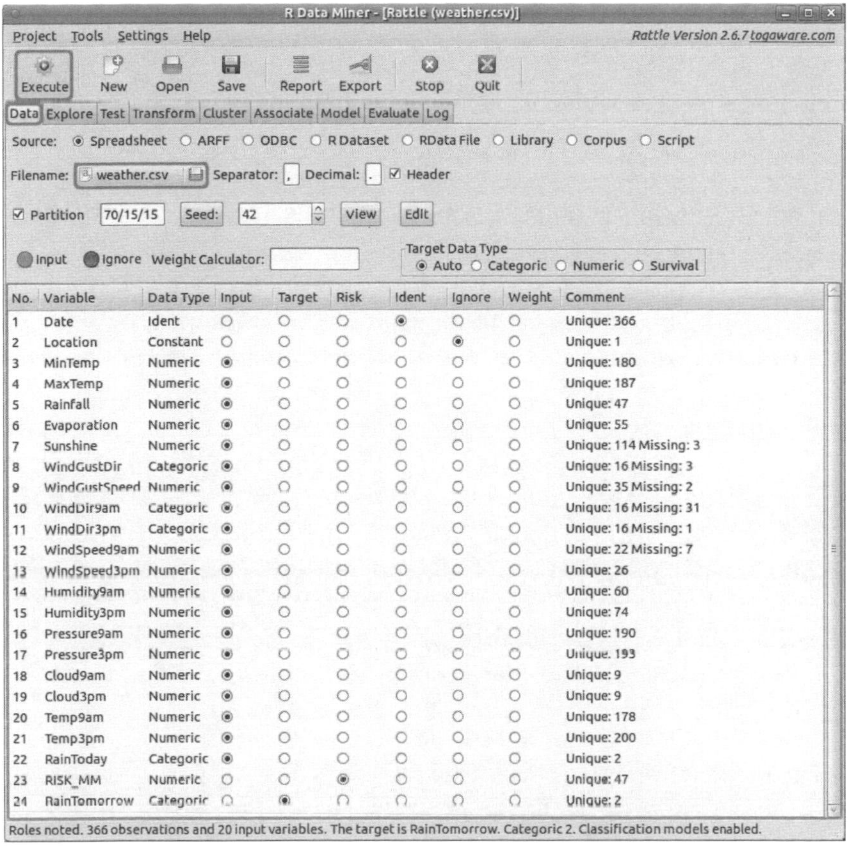

Figure 2.3: The sample weather.csv file has been loaded into memory as a dataset ready for mining. The dataset consists of 366 observations and 24 variables, as noted in the status bar. The first variable has a role other than the default Input role. Rattle uses heuristics to initialise the roles.

The file *weather.csv* will be loaded by default into Rattle as its dataset. Within R, a dataset is actually known as a *data frame*, and we will see this terminology frequently.

The dataset summary (Figure 2.3) provides a list of the variables, their data types, default roles, and other useful information. The types will generally be Numeric (if the data consists of numbers, like temperature, rainfall, and wind speed) or Categoric (if the data consists of characters from the alphabet, like the wind direction, which might be N or S, etc.), though we can also see an Ident (identifier). An Ident is often one of the variables (columns) in the data that uniquely identifies each observation (row) of the data. The Comments column includes general information like the number of unique (or distinct) values the variable has and how many observations have a missing value for a variable.

2.5   Building a Model

Using Rattle, we click the Model tab and are presented with the Model options (Figure 2.4). To build a decision tree model, one of the most common data mining models, click the Execute button (decision trees are the default). A textual representation of the model is shown in Figure 2.4.

The target variable (which stores the outcome we want to model or predict) is RainTomorrow, as we see in the Data tab window of Figure 2.3. Rattle automatically chose this variable as the target because it is the last variable in the data file and is a binary (i.e., two-valued) categoric. Using the *weather* dataset, our modelling task is to learn about the prospect of it raining tomorrow given what we know about today.

The textual presentation of the model in Figure 2.4 takes a little effort to understand and is further explained in Chapter 11. For now, we might click on the Draw button provided by Rattle to obtain the plot that we see in Figure 2.5. The plot provides a better idea of why it is called a decision **tree**. This is just a different way of representing the same model.

Clicking the Rules button will display a list of rules that are derived directly from the decision tree (we'll need to scroll the panel contained in the Model tab to see them). This is yet another way to represent the same model. The rules are listed here, and we explain them in detail next.

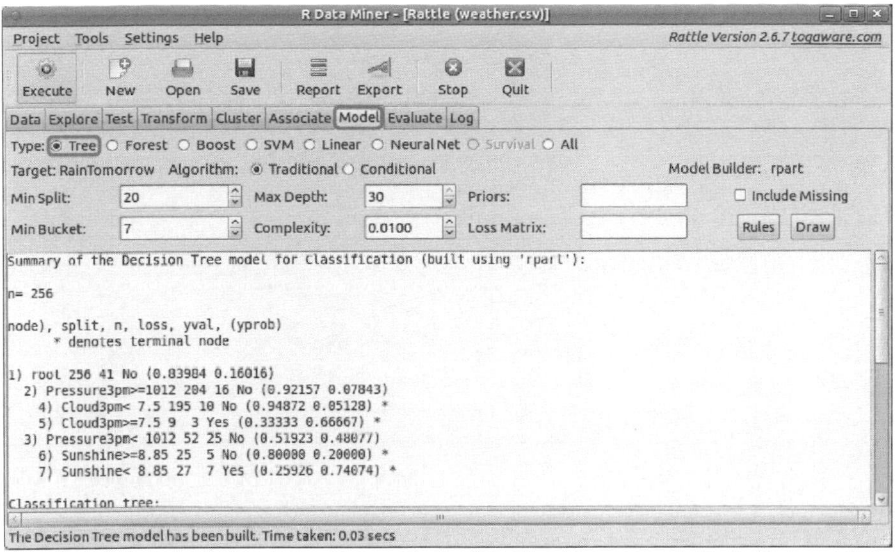

Figure 2.4: The *weather* dataset has been loaded, and a decision tree model has been built.

```
Rule number: 7 [RainTomorrow=Yes cover=27 (11%) prob=0.74]
   Pressure3pm< 1012
   Sunshine< 8.85

Rule number: 5 [RainTomorrow=Yes cover=9 (4%) prob=0.67]
   Pressure3pm>=1012
   Cloud3pm>=7.5

Rule number: 6 [RainTomorrow=No cover=25 (10%) prob=0.20]
   Pressure3pm< 1012
   Sunshine>=8.85

Rule number: 4 [RainTomorrow=No cover=195 (76%) prob=0.05]
   Pressure3pm>=1012
   Cloud3pm< 7.5
```

A well-recognised advantage of the decision tree representation for a model is that the paths through the decision tree can be interpreted as a collection of rules, as above. The rules are perhaps the more readable representation of the model. They are listed in the order of the prob-

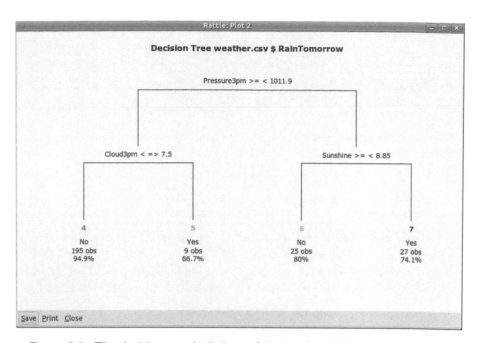

Figure 2.5: The decision tree built "out of the box" with Rattle. We traverse the tree by following the branches corresponding to the tests at each node. The > =< notation on the root (top) node indicates that we travel left if Pressure3pm is greater than 1011.9 and down the right branch if it is less than or equal to 1011.9. The <= > is similar, but reversed. The leaf nodes include a node number for reference, a decision of No or Yes to indicate whether it will RainTomorrow, the number of training observations, and the strength or confidence of the decision.

ability (prob) that we see listed with each rule. The interpretation of the probability will be explained in more detail in Chapter 11, but we provide an intuitive reading here.

Rule number 7 (which also corresponds to the "7)" in Figure 2.4 and leaf node number 7 in Figure 2.5) is the strongest rule predicting rain (having the highest probability for a Yes). We can read it as saying that if the atmospheric pressure (reduced to mean sea level) at 3 pm was less than 1012 hectopascals and the amount of sunshine today was less than 8.85 hours, then it seems there is a 74% chance of rain tomorrow ($yval = yes$ and $prob = 0.74$). That is to say that on most days when we have previously seen these conditions (as represented in the data) it has rained the following day.

Progressing down to the other end of the list of rules, we find the conditions under which it appears much less likely that there will be rain the following day. Rule number 4 has two conditions: the atmospheric pressure at 3 pm greater than or equal to 1012 hectopascals and cloud cover at 3 pm less than 7.5. When these conditions hold, the historic data tells us that it is unlikely to be raining tomorrow. In this particular case, it suggests only a 5% probability (`prob=0.05`) of rain tomorrow.

We now have our first model. We have data-mined our historic observations of weather to help provide some insight about the likelihood of it raining tomorrow.

2.6 Understanding Our Data

We have reviewed the modelling part of data mining above with very little attention to the data. A realistic data mining project, though, will precede modelling with quite an extensive exploration of data, in addition to understanding the business, understanding what data is available, and transforming such data into a form suitable for modelling. There is a lot more involved than just building a model. We look now at exploring our data to better understand it and to identify what we might want to do with it.

Rattle's Explore tab provides access to some common plots as well as extensive data exploration possibilities through **latticist** (Andrews, 2010) and **rggobi** (Lang et al., 2011). We will cover exploratory data analysis in detail in Chapters 5 and 6. We present here an initial flavour of exploratory data analysis.

One of the first things we might want to know is how the values of the target variable (`RainTomorrow`) are distributed. A histogram might be useful for this. The simplest way to create one is to go to the Data tab, click on the Input role for `RainTomorrow`, and click the Execute button. Then go to the Explore tab, choose the Distributions option, and then select Bar Plot for `RainTomorrow`. The plot of Figure 2.6 will be shown.

We can see from Figure 2.6 that the target variable is highly skewed. More than 80% of the days have no rain. This is typical of data mining, where even greater skewness is not uncommon. We need to be aware of the skewness, for example, in evaluating any models we build—a model that simply predicts that it never rains is going to be over 80% accurate, but pretty useless.

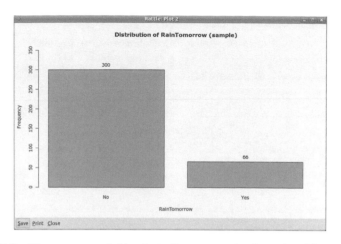

Figure 2.6: The target variable, RainTomorrow, is skewed, with Yes being quite underrepresented.

We can display other simple plots from the Explore tab by selecting the Distributions option. Under both the Box Plot and Histogram columns, select MaxTemp and Sunshine (as in Figure 2.7). Then click on Execute to display the plots in Figure 2.8. The plots begin to tell a story about the data. We sketch the story here, leaving the details to Chapter 5.

The top two plots are known as box-and-whisker plots. The top left plot tells us that the maximum temperature is generally higher the day before it rains (the plot above the x-axis label Yes) than before the days when it does not rain (above the No).

The top right plot suggests an even more dramatic skew for the amount of sunshine the day prior to the prediction. Generally we see that if there is less sunshine the day before, then the chance of rain (Yes) seems to be increased.

Both box plots also give another clue about the distribution of the values of the target variable. The width of the boxes in a box plot provides a visual indication of this distribution.

Each bottom plot overlays three separate plots that give further insight into the distribution of the observations. The three plots within each figure are a histogram (bars), a density plot (lines), and a rug plot (short spikes on the x-axis), each of which we now briefly describe.

The histogram has partitioned the numeric data into segments of equal width, showing the frequency for each segment. We see again that

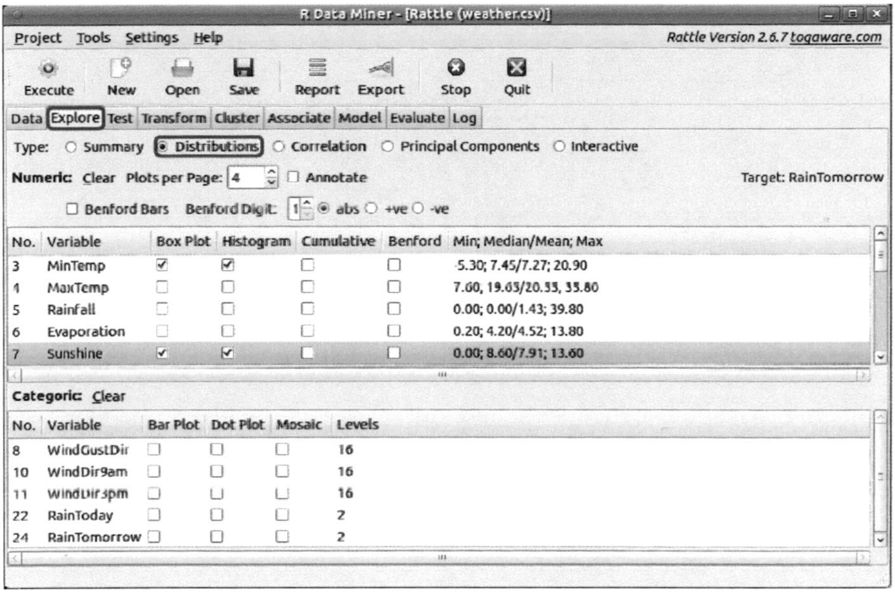

Figure 2.7: The *weather* dataset has been loaded and a decision tree model has been built.

sunshine (the bottom right) is quite skewed compared with the maximum temperature.

The density plots tend to convey a more accurate picture of the distribution of the data. Because the density plot is a simple line, we can also display the density plots for each of the target classes (Yes and No).

Along the x-axis is the rug plot. The short vertical lines represent actual observations. This can give us an idea of where any extreme values are, and the dense parts show where more of the observations lie.

These plots are useful in understanding the distribution of the numeric data. Rattle similarly provides a number of simple standard plots for categoric variables. A selection are shown in Figure 2.9. All three plots show a different view of the one variable, WindDir9am, as we now describe.

The top plot of Figure 2.9 shows a very simple bar chart, with bars corresponding to each of the levels (or values) of the categoric variable of interest (WindDir9am). The bar chart has been sorted from the overall most frequent to the overall least frequent categoric value. We note that each value of the variable (e.g., the value "SE," representing a wind direc-

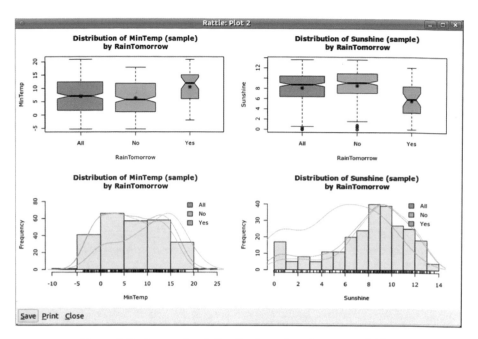

Figure 2.8: A sample of distribution plots for two variables.

tion of southeast) has three bars. The first bar is the overall frequency (i.e., the number of days) for which the wind direction at 9 am was from the southeast. The second and third bars show the breakdown for the values across the respective values of the categoric target variable (i.e., for No and Yes). We can see that the distribution within each wind direction differs between the three groups, some more than others. Recall that the three groups correspond to all observations (All), observations where it did not rain on the following day (No), and observations where it did (Yes).

The lower two plots show essentially the same information, in different forms. The bottom left plot is a dot plot. It is similar to the bar chart, on its side, and with dots representing the "top" of the bars. The breakdown into the levels of the target variable is compactly shown as dots within the same row.

The bottom right plot is a mosaic plot, with all bars having the same height. The relative frequencies between the values of WindDir9am are now indicated by the widths of the bars. Thus, SE is the widest bar, and WSW is the thinnest. The proportion between No and Yes within each bar

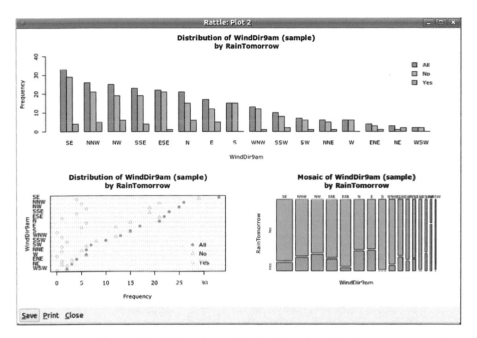

Figure 2.9: A sample of the three distribution plots for the one categoric variable.

is clearly shown.

A mosaic plot allows us to easily identify levels that have very different proportions associated with the levels of the target variable. We can see that a north wind direction has a higher proportion of observations where it rains the following day. That is, if there is a northerly wind today, then the chance of rain tomorrow seems to be increased.

These examples demonstrate that data visualisation (or exploratory data analysis) is a powerful tool for understanding our data—a picture is worth a thousand words. We actually learn quite a lot about our data even before we start to specifically model it. Many data miners begin to deliver significant benefits to their clients simply by providing such insights. We delve further into exploring data in Chapter 5.

2.7 Evaluating the Model: Confusion Matrix

We often begin a data mining project by exploring the data to gain our initial insights. In all likelihood, we then also transform and clean up our data in various ways. We have illustrated above how to then build our first model. It is now time to evaluate the performance or quality of the model.

Evaluation is a critical step in any data mining process, and one that is often left underdone. For the sake of getting started, we will look at a simple evaluation tool. The confusion matrix (also referred to as the *error matrix*) is a common mechanism for evaluating model performance.

In building our model we used a 70% subset of all of the available data. Figure 2.3 (page 27) shows the default sampling strategy of 70/15/15. We call the 70% sample the *training dataset*. The remainder is split equally into a *validation dataset* (15%) and a *testing dataset* (15%).

The validation dataset is used to test different parameter settings or different choices of variables whilst we are data mining. It is important to note that this dataset should not be used to provide any error estimations of the final results from data mining since it has been used as part of the process of building the model.

The testing dataset is *only* to be used to predict the unbiased error of the final results. It is important not to use this testing dataset in any way in building or even fine-tuning the models that we build. Otherwise, it no longer provides an unbiased estimate of the model performance.

The testing dataset and, whilst we are building models, the validation dataset, are used to test the performance of the models we build. This often involves calculating the model error rate. A confusion matrix simply compares the decisions made by the model with the actual decisions. This will provide us with an understanding of the level of accuracy of the model in terms of how well the model will perform on new, previously unseen, data.

Figure 2.10 shows the **Evaluate** tab with the **Error Matrix** (confusion matrix) using the **Testing** dataset for the **Tree** model that we have previously seen in Figures 2.4 and 2.5. Two tables are presented. The first lists the actual counts of observations and the second the percentages. We can observe that for 62% of the predictions the model correctly predicts that it won't rain (called the *true negatives*). That is, 35 days out of the 56 days are correctly predicted as not raining. Similarly, we see the model correctly predicts rain (called the *true positives*) on 18% of the days.

In terms of how correct the model is, we observe that it correctly predicts rain for 10 days out of the 15 days on which it actually does rain. This is a 67% accuracy in predicting rain. We call this the *true*

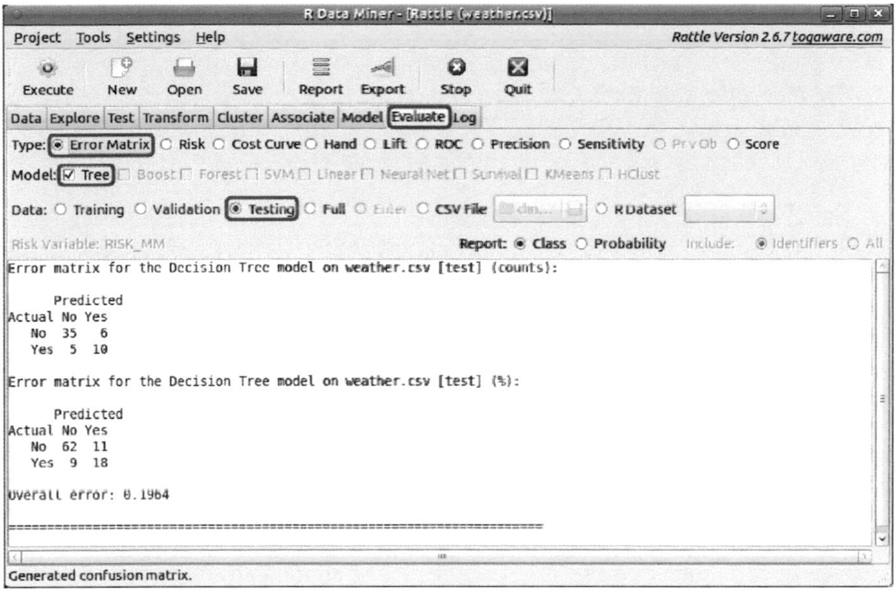

Figure 2.10: A confusion matrix applying the model to the testing dataset is displayed.

positive rate, but it is also known as the *recall* and the *sensitivity* of the model. Similarly, the *true negative rate* (also called the *specificity* of the model) is 85%.

We also see six days when we are expecting rain and none occurs (called the *false positives*). If we were using this model to help us decide whether to take an umbrella or raincoat with us on our travels tomorrow, then it is probably not a serious loss in this circumstance—we had to carry an umbrella without needing to use it. Perhaps more serious though is that there are five days when our model tells us there will be no rain yet it rains (called the *false negatives*). We might get inconveniently wet without our umbrella. The concepts of true and false positives and negatives will be further covered in Chapter 15.

The performance measure here tells us that we are going to get wet more often than we would like. This is an important issue—the fact that the different types of errors have different consequences for us. We'll also see more about this in Chapter 15.

It is useful to compare the performance as measured using the validation and testing datasets with the performance as measured using the training dataset. To do so, we can select the Validation and then the Training options (and for completeness the Full option) from the Data line of the Evaluate tab and then Execute each. The resulting performance will be reported. We reproduce all four here for comparison, including the count and the percentages.

Evaluation Using the Training Dataset:

Count		Predict		Percentage		Predict	
		No	Yes			No	Yes
Actual	No	205	10	Actual	No	80	4
	Yes	15	26		Yes	6	10

Evaluation Using the Validation Dataset:

Count		Predict		Percentage		Predict	
		No	Yes			No	Yes
Actual	No	39	5	Actual	No	72	9
	Yes	5	5		Yes	9	9

Evaluation Using the Testing Dataset:

Count		Predict		Percentage		Predict	
		No	Yes			No	Yes
Actual	No	35	6	Actual	No	62	11
	Yes	5	10		Yes	9	18

Evaluation Using the Full Dataset:

Count		Predict		Percentage		Predict	
		No	Yes			No	Yes
Actual	No	279	21	Actual	No	76	6
	Yes	25	41		Yes	7	11

We can see that there are fewer errors in the training dataset than in either the validation or testing datasets. That is not surprising since the tree was built using the training dataset, and so it should be more accurate on what it has already seen. This provides a hint as to why we do not validate our model on the training dataset—the evaluation will provide optimistic estimates of the performance of the model. By applying the model to the validation and testing datasets (which the model has not previously seen), we expect to obtain a more realistic estimate of the performance of the model on new data.

Notice that the overall accuracy from the training dataset is 90% (i.e., adding the diagonal percentages, 80% plus 10%), which is excellent. For the validation and testing datasets, it is around 80%. This is more likely how accurate the model will be longer-term as we apply it to new observations.

2.8 Interacting with **Rattle**

We have now stepped through some of the process of data mining. We have loaded some data, explored it, cleaned and transformed it, built a model, and evaluated the model. The model is now ready to be deployed. Of course, there is a lot more to what we have just done than what we have covered here. The remainder of the book provides much of these details. Before proceeding to the details, though, we might review how we interact with Rattle and R.

We have seen the Rattle interface throughout this chapter and we now introduce it more systematically. The interface is based on a set of tabs through which we progress as we work our way through a data mining project. For any tab, once we have set up the required information, we will click the Execute button to perform the actions. Take a moment to explore the interface a little. Notice the Help menu and that the help layout mimics the tab layout.

The Rattle interface is designed as a simple interface to a powerful suite of underlying tools for data mining. The general process is to step through each tab, left to right, performing the corresponding actions. For any tab, we configure the options and then click the Execute button (or F2) to perform the appropriate tasks. It is important to note that the tasks are **not** performed until the Execute button (or F2 or the Execute menu item under Tools) is clicked.

The Status Bar at the base of the window will indicate when the action is completed. Messages from R (e.g., error messages) may appear in the R Console from which Rattle was started. Since Rattle is a simple graphical interface sitting on top of R itself, it is important to remember that some errors encountered by R on loading the data (and in fact during any operation performed by Rattle) may be displayed in the R Console.

The R code that Rattle passes on to R to execute underneath the interface is recorded in the Log tab. This allows us to review the R commands that perform the corresponding data mining tasks. The R code snippets can be copied as text from the Log tab and pasted into the R Console from which Rattle is running, to be directly executed. This allows us to deploy Rattle for basic tasks yet still gives us the full power of R to be deployed as needed, perhaps through using more command options than are exposed through the Rattle interface. This also allows us the opportunity to export the whole session as an R script file.

The log serves as a record of the actions taken and allows those actions to be repeated directly and automatically through R itself at a later time. Simply select (to display) the Log tab and click on the Export button. This will export the log to a file that will have an R extension. We can choose to include or exclude the extensive comments provided in the log and to rename the internal Rattle variables (from "crs$" to a string of our own choosing).

We now traverse the main elements of the Rattle user interface, specifically the toolbar and menus. We begin with a basic concept—a project.

Projects

A project is a packaging of a dataset, variable selections, explorations, and models built from the data. Rattle allows projects to be saved for later resumption of the work or for sharing the data mining project with other users.

A project is typically saved to a file with a rattle extension. In fact, the file is a standard binary RData file used by R to store objects in a more compact binary form. Any R system can load such a file and hence have access to these objects, even without running Rattle.

Loading a rattle file into Rattle (using the Open button) will load that project into Rattle, restoring the data, models, and other displayed information related to the project, including the log and summary information. We can then resume our data mining from that point.

From a file system point of view, we can rename the files (as well as the filename extension, though that is not recommended) without impacting the project file itself—that is, the filename has no formal bearing on the contents, so use it to be descriptive. It is best to avoid spaces and unusual characters in the filenames.

Projects are opened and saved using the appropriate buttons on the toolbar or from the Project menu.

Toolbar

The most important button on the Toolbar (Figure 2.11) is the Execute button. All action is initiated with an Execute, often with a click of the Execute button. A keyboard shortcut for Execute is the F2 function key. A menu item for Execute is also available. It is worth repeating that the user interface paradigm used within Rattle is to set up the parameters on a tab and then Execute the tab.

Figure 2.11: The Rattle menu and toolbar.

The next few buttons on the Toolbar relate to the concept of a project within Rattle. Projects were discussed above.

Clicking on the New button will restore Rattle to its pristine startup state with no dataset loaded. This can be useful when a source dataset has been externally modified (external to Rattle and R). We might, for example, have manipulated our data in a spreadsheet or database program and re-exported the data to a CSV file. To reload this file into Rattle, if we have previously loaded it into the current Rattle session, we need to clear Rattle as with a click of the New button. We can then specify the filename and reload it.

The Report button will generate a formatted report based on the current tab. A number of report templates are provided with Rattle and will generate a document in the open standard ODT format, for the open source and open standards supporting LibreOffice. Whilst support for user-generated reports is limited, the log provides the necessary commands used to generate the ODT file. We can thus create our own ODT templates and apply them within the context of the current Rattle session.

The Export button is available to export various objects and entities from Rattle. Details are available together with the specific sections in the following chapters. The nature of the export depends on which tab is active and within the tab, which option is active. For example, if the Model tab is on display then Export will save the current model as PMML (the Predictive Modelling Markup Language—see Chapter 16). The Export button is not available for all tabs and options.

Menus

The menus (Figure 2.11) provide alternative access to many of the functions of the interface. A key point in introducing menus is that they can be navigated from the keyboard and contain keyboard shortcuts so that we can navigate more easily through Rattle using the keyboard.

The Project menu provides access to the Open and Save options for loading and saving projects from or to files. The Tools menu provides access to some of the other toolbar functions as well as access to specific tabs. The Settings menu allows us to control a number of optional characteristics of Rattle. This includes tooltips and the use of the more modern Cairo graphics device.

Extensive help is available through the Help menu. The structure of the menu follows that of the tabs of the main interface. On selecting a help topic, a brief text popup will display some basic information. Many of the popups then have the option of displaying further information, which will be displayed within a Web browser. This additional documentation comes directly from the documentation provided by R or the relevant R package.

Interacting with Plots

It is useful to know how we interact with plots in Rattle. Often we will generate plots and want to include them in our own reports. Plots are generated from various places within the Rattle interface.

Rattle optionally uses the Cairo device, which is a vector graphics engine for displaying high-quality graphic plots. If the Cairo device is not available within your installation, then Rattle resorts to the default window device for the operating system (x11() for GNU/Linux and window() for Microsoft Windows). The Settings menu also allows control of the choice of graphics device (allowing us to use the default by disabling support for Cairo). The Cairo device has a number of advantages, one being that it can be encapsulated within other windows, as is done with Rattle. This allows Rattle to provide some operating-system-independent functionality and a common interface. If we choose not to use the Cairo device, we will have the default devices, which still work just fine, but with less obvious functionality.

Figure 2.8 (page 34) shows a typical Rattle plot window. At the bottom of the window, we see a series of buttons that allow us to Save the plot to a file, to Print it, and Close it.

The Save button allows us to save the graphics to a file in one of the supported formats. The supported formats include pdf (for high-resolution pictures), png (for vector images and text), jpg (for colourful images), svg (for general scalable vector graphics), and, in Microsoft Windows, wmf (for Windows Metafile, Microsoft Windows-specific vector graphics). A popup will request the filename to save to. The default is to save in PDF format, saving to a file with the filename extension of .pdf. You can choose to save in the other formats simply by specifying the appropriate filename extension.

The Print button will send the plot to a printer. This requires the underlying R application to have been set up properly to access the required printer. This should be the case by default.

Once we are finished with the plot, we can click the Close button to shut down that particular plot window.

Keyboard Navigation

Keyboard navigation of the menus is usually initiated with the F10 function key. The keyboard arrow keys can then be used to navigate. Pressing the keyboard's Enter key will then select the highlighted menu item.

Judicious use of the keyboard (in particular, the arrow keys, the Tab and Shift-Tab keys, and the Enter key, together with F2 and F10) allows us to completely control Rattle from the keyboard if desired or required.

2.9 Interacting with R

R is a command line tool. We saw in Section 2.1 how to interact with R to start up Rattle. Essentially, R displays a prompt to indicate that it is waiting for us to issue a command. Two such commands are library() and rattle(). In this section, we introduce some basic concepts and commands for interacting with R directly.

Basic Functionality

Generally we instruct R to evaluate **functions**—a technical term used to describe mathematical objects that return a result. All functions in R return a result, and that result can be passed to other functions to do other things. This simple idea is actually a very powerful concept, allowing functions to do well what they are designed to do (like building a model) and pass on their output to other functions to do something with it (like formatting it for easy reading).

We saw in Section 2.1 two function calls, which we repeat below. The first was a call to the function library(), where we asked R to load **rattle**. We then started up Rattle with a call to the rattle() function:

```
> library(rattle)
> rattle()
```

Irrespective of the purpose of the function, for each function call we usually supply arguments that refine the behaviour of the function. We did that above in the call to library(), where the argument was rattle. Another simple example is to call dim() (dimensions) with the argument weather.

```
> dim(weather)

[1] 366   24
```

Here, weather is an *object* name. We can think of it simply as a reference to some object (something that contains data). The object in this case is the weather dataset as used in this chapter. It is organised as rows and columns. The dim() function reports the number of rows and columns.

If we type a name (e.g., either weather or dim) at the R prompt, R will respond by showing us the object. Typing weather (followed by pressing the Enter key) will result in the actual data. We will see all 366 rows of data scrolled on the screen. If we type dim and press Enter, we will see the definition of the function (which in this case is a primitive function coded into the core of R):

```
> dim

function (x)  .Primitive("dim")
```

A common mistake made by new users is to type a function name by itself (without arguments) and end up a little confused about the resulting output. To actually invoke the function, we need to supply the argument list, which may be an empty list. Thus, at a minimum, we add () to the function call on the command line:

```
> dim()

Error in dim: 0 arguments passed to 'dim' which requires 1
```

As we see, executing this function will generate an error message. We note that dim() actually needs one argument, and no arguments were passed to it. Some functions can be invoked with no arguments, as is the case for rattle().

The examples above illustrate how we will show our interaction with R. The "> " is R's prompt, and when we see that we know that R is waiting for commands. We type the string of characters dim(weather) as the command—in this case a call to the dim function. We then press the Enter key to send the command to R. R responds with the result from the function. In the case above, it returned the result [1] 366 24.

Technically, dim() returns a *vector* (a sequence of *elements* or values) of length 2. The [1] simply tells us that the first number we see from the vector (the 366) is the first element of the vector. The second element is 24.

The two numbers listed by R in the example above (i.e., the vector returned by dim()) are the number of rows and columns, respectively, in the *weather* dataset—that is, its dimensions.

For very long vectors, the list of the elements of the vector will be wrapped to fit across the screen, and each line will start with a number within square brackets to indicate what element of the vector we are up to. We can illustrate this with seq(), which generates a sequence of numbers:

```
> seq(1, 50)

 [1]  1  2  3  4  5  6  7  8  9 10 11 12 13 14 15 16 17 18
[19] 19 20 21 22 23 24 25 26 27 28 29 30 31 32 33 34 35 36
[37] 37 38 39 40 41 42 43 44 45 46 47 48 49 50
```

We saw above that we can view the actual data stored in an object by typing the name of the object (weather) at the command prompt.

Generally this will print too many lines (although only 366 in the case of the *weather* dataset). A useful pair of functions for inspecting our data are `head()` and `tail()`. These will list just the top and bottom six observations (or rows of data), by default, from the data frame, based on the order in which they appear there. Here we request, through the arguments to the function, to list the top two observations (and we also use indexing, described shortly, to list only the first nine variables):

```
> head(weather[1:9], 2)

          Date Location MinTemp MaxTemp Rainfall
1 2007-11-01 Canberra       8    24.3      0.0
2 2007-11-02 Canberra      14    26.9      3.6
  Evaporation Sunshine WindGustDir WindGustSpeed
1         3.4      6.3          NW            30
2         4.4      9.7         ENE            39
```

Similarly, we can request the bottom three rows of the dataset.

```
> tail(weather[1:9], 3)

            Date Location MinTemp MaxTemp Rainfall
364 2008-10-29 Canberra    12.5    19.9        0
365 2008-10-30 Canberra    12.5    26.9        0
366 2008-10-31 Canberra    12.3    30.2        0
    Evaporation Sunshine WindGustDir WindGustSpeed
364         8.4      5.3         ESE            43
365         5.0      7.1          NW            46
366         6.0     12.6          NW            78
```

The *weather* dataset is more complex than the simple vectors we have seen above. In fact, it is a special kind of list called a *data frame*, which is one of the most common data structures in R for storing our datasets. A data frame is essentially a list of columns. The *weather* dataset has 24 columns. For a data frame, each column is a vector, each of the same length.

If we only want to review certain rows or columns of the data frame, we can *index* the dataset name. Indexing simply uses square brackets to list the row numbers and column numbers that are of interest to us:

```
> weather[4:8, 2:4]

  Location MinTemp MaxTemp
4 Canberra    13.3    15.5
5 Canberra     7.6    16.1
6 Canberra     6.2    16.9
7 Canberra     6.1    18.2
8 Canberra     8.3    17.0
```

Notice the notation for a sequence of numbers. The string 4:8 is actually equivalent to a call to seq() with two arguments, 4 and 8. The function returns a vector containing the integers from 4 to 8. It's the same as listing them all and combining them using c():

```
> 4:8

[1] 4 5 6 7 8

> seq(4, 8)

[1] 4 5 6 7 8

> c(4, 5, 6, 7, 8)

[1] 4 5 6 7 8
```

Getting Help

It is important to know how we can learn more about using R. From the command line, we obtain help on commands by calling help():

```
> help(dim)
```

A shorthand is to precede the argument with a ? as in: ?dim. This is automatically converted into a call to help().

The help.search() function will search the documentation to list functions that may be of relevance to the topic we supply as an argument:

```
> help.search("dimensions")
```

The shorthand here is to precede the string with two question marks as in ??dimensions.

A third command for searching for help on a topic is RSiteSearch(). This will submit a query to the R project's search engine on the Internet:

```
> RSiteSearch("dimensions")
```

Quitting R

Recall that to exit from R, as we saw in Section 2.1, we issue q():

```
> q()
```

Our first session with R is now complete. The command line, as we have introduced here, is where we access the full power of R. But not everyone wants to learn and remember commands, so Rattle will get us started quite quickly into data mining, with only our minimal knowledge of the command line.

R and Rattle Interactions

Rattle generates R commands that are passed on through to R at various times during our interactions with Rattle. In particular, whenever the Execute button is clicked, Rattle constructs the appropriate R commands and then sends them off to R and awaits R's response.

We can also interact with R itself directly, and even interleave our interactions with Rattle and R. In Section 2.5, for example, we saw a decision tree model represented textually within Rattle's text view. The same can also be viewed in the R Console using print(). We can replicate that here once we have built the decision tree model as described in Section 2.5.

The R Console window is where we can enter R commands directly. We first need to make the window active, usually by clicking the mouse within that window. For the example below, we assume we have run Rattle on the *weather* dataset to build a decision tree as described in Section 2.5.

We can then type the print() command at the prompt. We see this in the code box below. The command itself consists of the name of an R function we wish to call on (print() in this case), followed by a list of *arguments* we pass to the function. The arguments provide information about what we want the function to do. The reference we see here, crs$rpart, identifies where the model itself has been saved internally by Rattle. The parameter digits= specifies the precision of the printed numbers. In this case we are choosing a single digit.

After typing the full command (including the function name and arguments) we then press the Enter key. This has the effect of passing the command to R. R will respond with the text exactly as shown below. The text starts with an indication of the number of observations (256). This is followed by the same textual presentation of the model we saw in Section 2.5.

```
> print(crs$rpart, digits=1)

n= 256

node), split, n, loss, yval, (yprob)
      * denotes terminal node

1) root 256 40 No (0.84 0.16)
  2) Pressure3pm>=1e+03 204 20 No (0.92 0.08)
    4) Cloud3pm< 8 195 10 No (0.95 0.05) *
    5) Cloud3pm>=8 9  3 Yes (0.33 0.67) *
  3) Pressure3pm< 1e+03 52 20 No (0.52 0.48)
    6) Sunshine>=9 25  5 No (0.80 0.20) *
    7) Sunshine< 9 27  7 Yes (0.26 0.74) *
```

Commands versus Functions

We have referred above to the R command line, where we enter commands to be executed. We also talked about functions that we type on the command line that make up the command to be executed. In this book, we will adopt a particular terminology around functions and commands, which we describe here.

In its true mathematical sense, a function is some operation that consumes some data and returns some result. Functions like dim(), seq(), and head(), as we have seen, do this. Functions might also have what we often call *side effects*—that is, they might do more than simply returning some result. In fact, the purpose of some functions is actually to perform some other action without necessarily returning a result. Such functions we will tend to call commands. The function rattle(), for example, does not return any result to the command line as such. Instead, its purpose is to start up the GUI and allow us to start data mining. Whilst rattle() is still a function, we will usually refer to it as a command rather than a function. The two terms can be used interchangeably.

Programming Styles for R

R is a programming language supporting different programming styles. We can use R to write programs that analyse data—we program the data analyses. Note that if we are only using Rattle, then we will not need to program directly. Nonetheless, for the programs we might write, we can take advantage of the numerous programming styles offered by R to develop code that analyses data in a consistent, simple, reusable, transparent, and error-free way.

Mistakenly, we are often trained to think that writing sentences in a programming language is primarily for the benefit of having a computer perform some activity for us. Instead, we should think of the task as really writing sentences that convey to other humans a story—a story about analysing our data. Coincidentally, we also want a computer to perform some activity.

Keeping this simple message in mind, whenever writing in R, helps to ensure we write in such a way that others can easily understand what we are doing and that we can also understand what we have done when we come back to it after six months or more.

Environments as Containers in R

For a particular project, we will usually analyse a collection of data, possibly transforming it and storing different bits of information about it. It is convenient to package all of our data and what we learn about it into some container, which we might save as a binary R object and reload more efficiently at a later time. We will use R's concept of an *environment* for this.

As a programming style, we can create a storage space and give it a name (i.e., it will look like a programming language variable) to act as a container. The container is an R environment and is initialised using new.env() (new environment). Here, we create a new environment and give it the name en:

```
> en <- new.env()
```

The object en now acts as a single container into which we can place all the relevant information associated with the dataset and that can also be shared amongst several models. We will store and access the relevant information from this container.

Data is placed into the container using the $ notation and the assignment operator, as we see in the following example:

```
> en$obs <- 4:8
> en$obs

[1] 4 5 6 7 8

> en$vars <- 2:4
> en$vars

[1] 2 3 4
```

The variables obs and vars are now contained within the environment referenced as en.

We can operate on variables within an environment without using the $ notation (which can become quite cumbersome) by wrapping the commands within evalq():

```
> evalq(
  {
    nobs <- length(obs)
    nvars <- length(vars)
  }, en)
> en$nobs

[1] 5

> en$nvars

[1] 3
```

The use of evalq() becomes most convenient when we have more than a couple of statements to write.

At any time, we can list the contents of the container using ls():

```
> ls(en)

[1] "nobs"   "nvars" "obs"    "vars"
```

Another useful function, provided by **gdata** (Warnes, 2011), is ll(), which provides a little more information:

```
> library(gdata)
> ll(en)

        Class KB
nobs    integer  0
nvars   integer  0
obs     integer  0
vars    integer  0
```

We can also convert the environment to a list using as.list():

```
> as.list(en)

$nvars
[1] 3

$nobs
[1] 5

$vars
[1] 2 3 4

$obs
[1] 4 5 6 7 8
```

By keeping all the data related to a project together, we can save and load the project through this one object. We also avoid "polluting" the global environment with lots of objects and losing track of what they all related to, possibly confusing ourselves and others.

We can now also quite easily use the same variable names, but within different containers. Then, when we write scripts to build models, for example, often we will be able to use exactly the same scripts, changing only the name of the container. This encourages the reuse of our code and promotes efficiencies.

This approach is also sympathetic to the concept of object-oriented programming. The container is a basic "object" in the object-oriented programming context.

We will use this approach of encapsulating all of our data and information within a container when we start building models. The following provides the basic template:

```
> library(rpart)
> weatherDS <- new.env()
> evalq({
    data <- weather
    nobs <- nrow(data)
    vars <- c(2:22, 24)
    form <- formula(RainTomorrow ~ .)
    target <- all.vars(form)[1]
    train <- sample(nobs, 0.7*nobs)
  }, weatherDS)
> weatherRPART <- new.env(parent=weatherDS)
> evalq({
    model <- rpart(formula=form, data=data[train, vars])
    predictions <- predict(model, data[-train, vars])
  }, weatherRPART)
```

Here we have created two containers, one for the data and the other for
the model. The model container (weatherRPART) has as its parent the
data container (weatherDS), which is achieved by specifying the parent=
argument. This makes the variables defined in the data container available within the model container.

To save a container to a file for use at a later time, or to document
stages within the data mining project, use save():

```
> save(weatherDS, file="weatherDS.Rdata")
```

It can later be loaded using load():

```
> load("weatherDS.Rdata")
```

It can at times become tiresome to be wrapping our code up within
a container. Whilst we retain the discipline of using containers we can
also quickly interact with the variables in a container without having to
specify the container each time. WE use attach and detach to add a
container into the so called search path used by R to find variables. Thus
we could do something like the following:

```
> attach(weatherRPART)
> print(model)
> detach(weatherRPART)
```

However, creating new variables to store within the environment will not work in the same way. Thus:

```
> attach(weatherRPART)
> new.model <- model
> detach(weatherRPART)
```

does not place the variable `new.model` into the `weatherRPART` environment. Instead it goes into the global environment.

A convenient feature, particularly with the layout used within the `evalq()` examples above and generally throughout the book, is that we could ignore the string that starts a block of code (which is the line containing "`evalq({`") and the string that ends a block of code (which is the line containing "`}, weatherDS)`") and simply copy-and-paste the other commands directly into the R console. The variables (`data`, `nobs`, etc.) are then created in the global environment, and nothing special is needed to access them. This is useful for quickly testing out ideas, for example, and is provided as a choice if you prefer not to use the container concept yourself. Containers do, however, provide useful benefits.

Rattle uses containers internally to collect together the data it needs. The Rattle container is called `crs` (the current rattle store). Once a dataset is loaded into Rattle, for example, it is stored as `crs$dataset`. We saw `crs$rpart` above as referring to the decision tree we built above.

2.10   Summary

In this chapter, we have become familiar with the Rattle interface for data mining with R. We have also built our first data mining model, albeit using an already prepared dataset. We have also introduced some of the basics of interacting with the R language.

We are now ready to delve into the details of data mining. Each of the following chapters will cover a specific aspect of the data mining process and illustrate how this is accomplished within Rattle and then further extended with direct coding in R.

Before proceeding, it is advisable to review Chapter 1 as an introduction to the overall data mining process if you have not already done so.

2.11   Command Summary

This chapter has referenced the following R packages, commands, functions, and datasets:

`<-`	function	Assign a value into a named reference.
`c()`	function	Concatenate values into a vector.
`dim()`	function	Return the dimensions of a dataset.
`evalq()`	function	Access the environment for storing data.
`head()`	function	Return the first few rows of a dataset.
`help()`	command	Display help for a specific function.
`help.search()`	command	Search for help on a specific topic.
latticist	package	Interactive visualisation of data.
`library()`	command	Load a package into the R library.
`ll()`	function	Longer list of an environment.
`load()`	command	Load R objects from a file.
`ls()`	function	List the contents of an environment.
`new.env()`	function	Create a new object to store data.
`nrow()`	function	Number of rows in a dataset.
`print()`	command	Display representation of an R object.
`q()`	command	Quit from R.
`R`	shell	Start up the R statistical environment.
`rattle()`	command	Start the Rattle GUI.
rggobi	package	Interactive visualisation of data.
`rpart()`	function	Build a decision tree predictive model.
rpart	package	Provides decision tree functions.
`RSiteSearch()`	command	Search the R Web site for help.
`sample()`	function	Random selection of its first argument.
`save()`	command	Save R objects into a file.
`seq()`	function	Return a sequence of numbers.
`table()`	function	Make a table from some variables.
`tail()`	function	Return the last few rows of a dataset.
weather	dataset	Sample dataset from **rattle**.
`window()`	command	Open a new plot in Microsoft Windows.
`x11()`	command	Open a new plot in Unix/Linux.

Chapter 3

Working with Data

Data is the starting point for all data mining—without it there is nothing to mine. In today's world, there is certainly no shortage of data, but turning that data into information, knowledge, and, eventually, wisdom is not a simple matter. We often think of data as being numbers or categories. But data can also be text, images, videos, and sounds. Data mining generally only deals with numbers and categories. Often, the other forms of data can be mapped into numbers and categories if we wish to analyse them using the approaches we present here.

Whilst data abounds in our modern era, we still need to scout around to obtain the data we need. Many of today's organisations maintain massive warehouses of data. This provides a fertile ground for sourcing data but also an extensive headache for us in navigating through a massive landscape.

An early step in a data mining project is to gather all the required data together. This seemingly simple task can be a significant burden on the budgeted resources for data mining, perhaps consuming up to 70–90% of the elapsed time of a project. It should not be underestimated.

When bringing data together, a number of issues need to be considered. These include the provenance (source and purpose) and quality (accuracy and reliability) of the data. Data collected for different purposes may well store different information in confusingly similar ways. Also, some data requires appropriate permission for its use, and the privacy of anyone the data relates to needs to be considered. Time spent at this stage getting to know your data will be time well spent.

In this chapter, we introduce data, starting with the language we use to describe and talk about data.

3.1 Data Nomenclature

Data miners have a plethora of terminology, often using many different terms to describe the same concept. A lot of this confusion of terminology is due to the history of data mining, with its roots in many different disciplines, including databases, machine learning, and statistics. Throughout this book, we will use a consistent and generally accepted nomenclature, which we introduce here.

We refer to a collection of data as a **dataset**. This might be called in mathematical terms a *matrix* or in database terms a *table*. Figure 3.1 illustrates a dataset annotated with our chosen nomenclature.

We often view a dataset as consisting of rows, which we refer to as **observations**, and those observations are recorded in terms of **variables**, which form the columns of the dataset. Observations are also known as *entities*, *rows*, *records*, and *objects*. Variables are also known as *fields*, *columns*, *attributes*, *characteristics*, and *features*. The **dimension** of a dataset refers to the number of observations (rows) and the number of variables (columns).

Variables can serve different roles: as **input variables** or **output variables**. Input variables are *measured* or *preset* data items. They might also be known as *predictors, covariates, independent variables, observed variables*, and *descriptive variables*. An output variable may be identified in the data. These are variables that are often "influenced" by the input variables. They might also be known as *target, response*, or *dependent variables*. In data mining, we often build models to predict the output variables in terms of the input variables. Early on in a data mining project, we may not know for sure which variables, if any, are output variables. For some data mining tasks (e.g., clustering), we might not have any output variables.

Some variables may only serve to uniquely identify the observations. Common examples include social security and other such government identity numbers. Even the date may be a unique identifier for particular observations. We refer to such variables as **identifiers**. Identifiers are not normally used in modelling, particularly those that are essentially randomly generated.

Variables can store different types of data. The values might be the names or the qualities of objects, represented as character strings. Or the values may be quantitative and thereby represented numerically. At a high level we often only need to distinguish these two broad types of data, as we do here.

	Date	Temp	Wind Dir.	Evap	Rain?
	10 Dec	23	NNE	10.4	Y
	25 Jan	25	E	6.8	Y
	02 Apr	22	SSW	3.6	N
	08 May	17		4.4	N
	10 May	21	NW	2.4	N
	04 Jun	13	SE	0.2	Y
	04 Jul	10	SSW	1.8	N
	01 Aug	9	NW	2.6	N
	07 Aug	6	SE	3.0	Y

Variables →   (columns: Date, Temp, Wind Dir., Evap, Rain?)
Observations →

Column types: Numeric (Temp), Categoric Numeric (Wind Dir.), Numeric (Evap), Categoric (Rain?)

Identifier   Input   Output

Figure 3.1: A simple dataset showing the nomenclature used. Each column is a *variable* and each row is an *observation*.

A **categoric variable**[1] is one that takes on a single value, for a particular observation, from a fixed set of possible values. Examples include eye colour (with possible values including blue, green, and brown), age group (with possible values young, middle age, and old), and rain tomorrow (with only two possible values, Yes and No). Categoric variables are always *discrete* (i.e., they can only take on specific values).

Categoric variables like eye colour are also known as *nominal variables*, *qualitative variables*, or *factors*. The possible values have no order to them. That is, blue is no less than or greater than green.

On the other hand, categoric variables like age group are also known as *ordinal variables*. The possible values have a natural order to them, so that young is in some sense less than middle age, which in turn is less than old.

---

[1]We use the terms *categoric* rather than *categorical* and *numeric* rather than *numerical*.

A categoric variable like `rain tomorrow`, having only two possible values, is also known as a *binary variable*.

A **numeric variable** has values that are integers or real numbers, such as a person's age or weight or their income or bank balance. Numeric variables are also known as *quantitative variables*. Numeric variables can be *discrete* (integers) or *continuous* (real).

A **dataset** (or, in particular, different randomly chosen subsets of a dataset) can have different roles. For building predictive models, for example, we often partition a dataset into three independent datasets: a **training dataset**, a **validation dataset**, and a **testing dataset**. The partitioning is done randomly to ensure each dataset is representative of the whole collection of observations. Typical splits might be 40/30/30 or 70/15/15. A validation dataset is also known as a design dataset (since it assists in the design of the model).

We build our model using the training dataset. The validation dataset is used to assess the model's performance. This will lead us to tune the model, perhaps through setting different model parameters. Once we are satisfied with the model, we assess its expected performance into the future using the testing dataset.

It is important to understand the significance of the testing dataset. This dataset must be a so-called *holdout* or *out-of-sample* dataset. It consists of randomly selected observations from the full dataset that are not used in any way in the building of the model. That is, it contains no observations in common with the training or validation datasets. This is important in relation to ensuring we obtain an unbiased estimate of the true performance of a model on new, previously unseen observations.

We can summarise our generic nomenclature, in one sentence, as:

> A **dataset** consists of **observations** recorded using **variables**, which consist of a mixture of **input variables** and **output variables**, either of which may be **categoric** or **numeric**.

Having introduced our generic nomenclature, we also need to relate the same concepts to how they are implemented in an actual system, like R. We do so, briefly, here.

R has the concept of a *data frame* to represent a dataset. A data frame is, technically, a *list* of variables. Each variable in the list represents a column of data—a variable stores a collection of data items that are all of the same type. For example, this might be a collection of integers recording the ages of clients. Technically, R refers to what we call a variable within a dataset as a *vector*.

Each variable will record the same number of data items, and thus we can picture the dataset as a rectangular matrix, as we illustrated in Figure 3.1. A data frame is much like a table in a database or a page in a spreadsheet. It consists of rows, which we have called *observations*, and columns, which we have called *variables*.

3.2   Sourcing Data for Mining

To start a data mining project, we must first recognise and understand the problem to tackle. Whilst that might be quite obvious, there are subtleties we need to address, as discussed in Chapter 1. We also need data—again, somewhat obvious. As we suggested above, though, sourcing our data is usually not a trivial matter. We discuss the general data issue here before we delve into some technical aspects of data.

In an ideal world, the data we require for data mining will be nicely stored in a data warehouse or a database, or perhaps a spreadsheet. However, we live in a less than ideal world. Data is stored in many different forms and on many different systems, with many different meanings. Data is everywhere, for sure, but we need to find it, understand it, and bring it together.

Over the years, organisations have implemented well-managed data warehouse systems. They serve as the organisation-wide repository of data. It is true, though that, despite this, data will always spring up outside of the data warehouse, and will have none of the careful controls that surround the data warehouse with regard to data provenance and data quality. Eventually the organisation's data custodians will recapture the useful new "cottage industry" repositories into the data warehouse and the cycle of new "cottage industries" will begin once again. We will always face the challenge of finding data from many sources within an organisation.

An organisation's data is often not the only data we access within a data mining project. Data can be sourced from outside the organisation. This could include data publicly available, commercially collected, or legislatively obtained. The data will be in a variety of formats and of varying quality. An early task for us is to assess whether the data will be useful for the business problem and how we will bring the new data together with our other data. We delve further into understanding the data in Chapter 5. We consider data quality now.

3.3 Data Quality

No real-world data is perfectly collected. Despite the amount of effort organisations put into ensuring the quality of the data they collect, errors will always occur. We need to understand issues relating to, for example, consistency, accuracy, completeness, interpretability, accessibility, and timeliness.

It is important that we recognise and understand that our data will be of varying quality. We need to treat (i.e., transform) our data appropriately and be aware of the limitations (uncertainties) of any analysis we perform on it. Chapter 7 covers many aspects of data quality and how we can work towards improving the quality of our available data. Below we summarise some of the issues.

In the past, much data was entered by data entry staff working from forms or directly in conversation with clients. Different data entry staff often interpret different data fields (variables) differently. Such inconsistencies might include using different formats for dates or recording expenses in different currencies in the same field, with no information to identify the currency.

Often in the collection of data some data is more carefully (or accurately) collected than other data. For bank transactions, for example, the dollar amounts must be very accurate. The precise spelling of a person's name or address might not need to be quite so accurate. Where the data must be accurate, extra resources will be made available to ensure data quality. Where accuracy is less critical, resources might be saved. In analysing data, it is important to understand these aspects of accuracy.

Related to accuracy is the issue of completeness. Some less important data might only be optionally collected, and thus we end up with much missing data in our datasets. Alternatively, some data might be hard to collect, and so for some observations it will be missing. When analysing data, we need to understand the reasons for missing data and deal with the data appropriately. We cover this in detail in Chapter 7.

Another major issue faced by the data miner is the interpretation of the data. Having a thorough understanding of the meaning of the data is critical. Knowing that height is measured in feet or in meters will make a difference to the analysis. We might find that some data was entered as feet and other data as meters (the consistency problem). We might have dollar amounts over many years, and our analysis might need to interpret the amounts in terms of their relative present-day value. Codes are also often used, and we need to understand what each code means and how different codes relate to each other. As the data ages, the meaning of the different variables will often change or be altogether lost. We need to understand and deal with this.

The accessibility of the right data for analysis will often also be an issue. A typical process in data collection involves checking for obvious data errors in the data supplied and correcting those errors. In collecting tax return data from taxpayers, for example, basic checks will be performed to ensure the data appears correct (e.g., checking for mistakes that enter data as 3450 to mean $3450, whereas it was meant to be $34.50). Sometimes the checks might involve discussing the data with its supplier and modifying it appropriately. Often it is this "cleaner" data that is stored on the system rather than the original data supplied. The original data is often archived, but often it is such data that we actually need for the analysis—we want to analyse the data as supplied originally. Accessing archived data is often problematic.

Accessing the most recent data can sometimes be a challenge. In an online data processing environment, where the key measure of performance is the turnaround time of the transaction, providing other systems with access to the data in a timely manner can be a problem. In many environments, the data can only be accessed after a sometimes complex extract/transform/load (ETL) process. This can mean that the data may only be available after a day or so, which may present challenges for its timely analysis. Often, business processes need to be changed so that more timely access is possible.

3.4   Data Matching

In collecting data from multiple sources, we end up with a major problem in that we need to match observations from one dataset with those from another dataset. That is, we need to identify the same entities (e.g., people or companies) from different data sources. These different sources could be, for example, patient medical data from a doctor and from a hospital. The doctor's data might contain information about the patients' general visits, basic test results, diagnoses, and prescriptions. The doctor might have a unique number to identify his or her own patients, as well as their names, dates of birth, and addresses. A hospital will also record data about patients that are admitted, including their reason for admission, treatment plan, and medications. The hospital will probably have its own unique number to identify each patient, as well as the patient's name, date and place of birth, and address.

The process of *data matching* might be as simple as joining two datasets together based on shared identifiers that are used in each of the two databases. If the doctor and the hospital share the same unique numbers to identify the patients, then the data matching process is simplified.

However, the data matching task is usually much more complex. Data matching often involves, for example, matching of names, addresses, and dates and places of birth, all of which will have inaccuracies and alternatives for the same thing. The data entered at a doctor's consulting rooms will in general be entered by a different receptionist on a different day from the data entered on admission at a hospital where surgery might be performed.

It is not uncommon to find, even within a single database, one person's name recorded differently, let alone when dealing with data from very different sources. One data source might identify "John L. Smith," and another might identify the person as "J.L. Smith," and a third might have an error or two but identify the person as "Jon Leslie Smyth."

The task of data matching is to bring different data sources together in a reliable and supportable manner so that we have the right data about the right person. An idea that can improve data matching quality is that of a trusted data matching bureau. Many data matching bureaus within organisations almost start each new data matching effort from scratch. However, over time there is the opportunity to build up a data matching database that records relevant information about all previous data matches.

Under this scenario, each time a new data matching effort is undertaken, the identities within this database, and their associated information, are used to improve the new data matching. Importantly, the results of the new data matching feed back into the data matching database to improve the quality of the matched entities and thus even improve previously matched data.

Data matching is quite an extensive topic in itself and worth a separate book. A number of commercially available tools assist with the basic task. The open source Febrl[2] system also provides data matching capabilities. They all aim to identify the same entity in all of the data sources.

3.5 Data Warehousing

The process of bringing data together into one unified and carefully managed repository is referred to as *data warehousing*—the analogy being with a large building used for the storage of goods. What we store in our warehouse is data. Data warehouses were topical in the 1990s and primarily vendor driven, servicing a real opportunity to get on top of managing data. Inmon (1996) provides a detailed introduction.

We can view a data warehouse as a large database management system. It is designed to integrate data from many different sources and to support analysis for different objectives. In any organisation, the data warehouse can be the foundation for business intelligence, providing a single, integrated source of data for the whole organisation.

Typically, a data warehouse will contain data collected together from multiple sources but focussed around the function of an organisation. The data sources will often be *operational systems* (such as transaction processing systems) that run the day-to-day functions of the organisation. In banking, for example, the transaction processing systems include ATMs and EFTPOS machines, which are today most pervasive. Transaction processing systems collect data that gets uploaded to the data warehouse on a regular basis (e.g., daily, but perhaps even more frequently).

Well-organised data warehouses, at least from the point of view of data mining, will also be nonvolatile. The data stored in our data warehouses will capture data regularly, and older data is not removed. Even when an update to data is to correct existing data items, such data must be maintained, creating a massive repository of historic data that can be used to carefully track changes over time.

Consider the case of tax returns held by our various revenue authorities. Many corrections are made to individual tax returns over time. When a tax return is filed, a number of checks for accuracy may result in

---

[2]http://sourceforge.net/projects/febrl/.

simple changes (e.g., correcting a misspelled address). Further changes might be made at a later time as a taxpayer corrects data originally supplied. Changes might also be the result of audits leading to corrections made by the revenue authority, or a taxpayer may notify the authority of a change in address.

Keeping the history of data changes is essential for data mining. It may be quite significant, from a fraud point of view, that a number of clients in a short period of time change their details in a common way. Similarly, it might be significant, from the point of view of understanding client behaviour, that a client has had ten different addresses in the past 12 months. It might be of interest that a taxpayer always files his or her tax return on time each year, and then makes the same two adjustments subsequently, each year. All of this historic data is important in building a picture of the entities we are interested in. Whilst the operational systems may only store data for one or two months before it is archived, having this data accessible for many years within a data warehouse for data mining is important.

In building a data warehouse, much effort goes into how the data warehouse is structured. It must be designed to facilitate the queries that operate on a large proportion of data. A careful design that exposes all of the data to those who require it will aid in the data mining process.

Data warehouses quickly become unwieldly as more data is collected. This often leads to the development of specific *data marts*, which can be thought of as creating a tuned subset of the data warehouse for specific purposes. An organisation, for example, may have a finance data mart, a marketing data mart, and a sales data mart. Each data mart will draw its information from various other data collected in the warehouse. Different data sources within the warehouse will be shared by different data marts and present the data in different ways.

A crucial aspect of a data warehouse (and any data storage, in fact) is the maintenance of information about the data—so-called **metadata**. Metadata helps make the data understandable and thereby useful. We might talk about two types of metadata: **technical metadata** and **business metadata**.

Technical metadata captures data about the operational systems from which the data was obtained, how it was extracted from the source systems, how it was transformed, how it was loaded into the warehouse, where it is stored within the warehouse, and its structure as stored in the warehouse.

The actual process of extracting, transforming, and then loading data is often referred to as ETL (extract, transform, load). Many vendors provide ETL tools, and there is also extensive capability for automating ETL using open source software, including R.

The business metadata, on the other hand, provides the information that is useful in understanding the data. It will include descriptions of the variables contained in the data and measures of their data quality. It can also include who "owns" the data, who has access to it, the cost of accessing it, when it was last updated, how frequently it is updated, and how it is used operationally.

Before data mining became a widely adopted technology, the data warehouse supported analyses through business intelligence (BI) technology. The simplest analyses build reports that aggregate the data within a warehouse in many different ways. Through this technology, an organisation is able to ensure its executives are aware of its activities. On-line, analytic processing (OLAP) within the BI technology supports user-driven and multidimensional analyses of the data contained within the warehouse. Extending the concept of a human-driven and generally manual analysis of data, as in business intelligence, data mining provides a data-driven approach to the analysis of the data.

Ideally, the data warehouse is the primary data source for data mining. Integrating data from multiple sources, the data warehouse should contain an extensive resource that captures all of the activity of an organisation. Also, ideally, the data will be consistent, of high quality, and documented with very good metadata. If all that is true, the data mining will be quite straightforward. Rarely is this true. Nonetheless, mining data from the data warehouse can significantly reduce the time for preparing it and sharing the data across many data mining and reporting projects.

Data warehouses will often be accessed through the common *structured query language* (SQL). Our data will usually be spread across multiple locations within the warehouse, and SQL queries will be used to bring them together. Some basic familiarity with SQL will be useful as we extract our data. Otherwise we will need to ensure we have ready access to the skills of a data analyst to extract the data for us.

3.6   Interacting with Data Using R

Once we have scouted for data, matched common entities, and brought the data together, we need to structure the data into a form suitable for data mining. More specifically, we need to structure the data to suit the data mining tool we are intending to use. In our case, this involves putting the data into a form that allows it to be easily loaded into R, using Rattle, where we will then explore, test, and transform it in various ways in preparation for mining.

Once we have loaded a dataset into Rattle, through one of the mechanisms we introduce in Chapter 4 (or directly through R itself), we may want to modify the data, clean it, and transform it into the structures we require. We may already be familiar with a variety of tools for dealing with data (like SQL or a spreadsheet). These tools may be quite adequate for the manipulations we need to undertake. We can easily prepare the data with them and then load it into Rattle when ready. But R itself is also a very powerful data manipulation language.

Much of R's capabilities for data management are covered in other books, including those of Spector (2008), Muenchen (2008), and Chambers (2008). Rattle provides access to some data cleaning operations under the Transform tab, as covered in Chapter 7. We provide here elementary instruction in using R itself for a limited set of manipulations that are typical in preparing data for data mining. We do not necessarily cover the details nor provide the systematic coverage of R available through other means.

One of the most basic operations is accessing the data within a dataset. We *index* a dataset using the notation of square brackets, and within the square brackets we identify the index of the observations and the variables we are interested in, separating them with a comma. We briefly saw this previously in Section 2.9.

Using the same *weather* dataset as in Chapter 2 (available from **rattle**, which we can load into R's library()), we can access observations 100 to 105 and variables 5 to 6 by *indexing* the dataset. If either index (observations or variables) is left empty, then the result will be all observations or all variables, respectively, rather than just a subset of them. Using dim() to report on the resulting size (dimensions) of the dataset, we can see the effect of the indexing:

```
> library(rattle)
> weather[100:105, 5:6]

     Rainfall Evaporation
100     16.2          5.4
101      0.0          4.0
102      0.0          5.8
103      0.0          5.0
104      4.4          6.6
105     11.0          3.2

> dim(weather)

[1] 366  24

> dim(weather[100:105, 5:6])

[1] 6 2

> dim(weather[100:105,])

[1]   6 24

> dim(weather[,5:6])

[1] 366   2

> dim(weather[5:6])

[1] 366   2

> dim(weather[,])

[1] 366  24
```

Note that the notation 100:105 is actually shorthand for a call to seq(), which generates a list of numbers. Another way to generate a list of numbers is to use c() (for combine) and list each of the numbers explicitly. These expressions can replace the "100:105" in the example above to have the same effect. We can see this in the following code block.

```
> 100:105

[1] 100 101 102 103 104 105

> seq(100, 105)

[1] 100 101 102 103 104 105

> c(100, 101, 102, 103, 104, 105)

[1] 100 101 102 103 104 105
```

Variables can be referred to by their position number, as above, or by the variable name. In the following example, we extract six observations of just two variables. Note the use of the `vars` object to list the variables of interest and then from that index the dataset.

```
> vars <- c("Evaporation", "Sunshine")
> weather[100:105, vars]

    Evaporation Sunshine
100         5.4      5.6
101         4.0      8.9
102         5.8      9.6
103         5.0     10.7
104         6.6      5.9
105         3.2      0.4
```

We can list the variable names contained within a dataset using `names()`:

```
> head(names(weather))

[1] "Date"       "Location"    "MinTemp"
[4] "MaxTemp"    "Rainfall"    "Evaporation"
```

In this example we list only the first six names, making use of `head()`. This example also illustrates the "functional" nature of R. Notice how we directly feed the output of one function (`names()`) into another function (`head()`).

We could also use indexing to achieve the same result:

```
> names(weather)[1:6]

[1] "Date"       "Location"    "MinTemp"
[4] "MaxTemp"    "Rainfall"    "Evaporation"
```

When we index a dataset with single brackets, as in `weather[2]` or `weather[4:7]`, we retrieve a "subset" of the dataset—specifically, we retrieve a subset of the variables. The result itself is another dataset, even if it contains just a single variable. Compare this with `weather[[2]]`, which returns the actual values of the variable. The differences may appear subtle, but as we gain experience with R, they become important. We do not dwell on this here, though.

```
> head(weather[2])

  Location
1 Canberra
2 Canberra
3 Canberra
4 Canberra
5 Canberra
6 Canberra

> head(weather[[2]])

[1] Canberra Canberra Canberra Canberra Canberra Canberra
46 Levels: Adelaide Albany Albury ... Woomera
```

We can use the $ notation to access specific variables within a dataset. The expression `weather$MinTemp` refers to the `MinTemp` variable of the *weather* dataset:

```
> head(weather$MinTemp)

[1]  8.0 14.0 13.7 13.3  7.6  6.2
```

3.7 Documenting the Data

The *weather* dataset, for example, though very small in the number of observations, is somewhat typical of data mining. We have obtained the dataset from a known source and have processed it to build a dataset ready for our data mining. To do this, we've had to research the meaning of the variables and read about any idiosyncrasies associated with the collection of the data. Such information needs to be captured in a data mining report. The report should record where our data has come from, our understanding of its integrity, and the meaning of the variables. This information will come from a variety of sources and usually from multiple domain experts. We need to understand and document the provenance of the data: how it was collected, who collected it, and how they understood what they were collecting.

The following summary will be useful. It is obtained from processing the output from the str(). That output, which is normally only displayed in the console, is first captured into a variable using capture.output():

```
> sw <- capture.output(str(weather, vec.len=1))
> cat(sw[1])

'data.frame':        366 obs. of   24 variables:
```

The output is then processed to add a variable number and appropriately fit the page. The processing first uses sprintf() to generate a list of variable numbers, each number stored as a string of width 2 ("%2d"):

```
> swa <- sprintf("%2d", 1:length(sw[-1]))
```

Each number is then pasted to each line of the output, collapsing the separate lines to form one long string with a new line ("\n") separating each line:

```
> swa <- paste(swa, sw[-1], sep="", collapse="\n")
```

The gsub() function is then used to truncate lines that are too long by substituting a particular pattern of dots and digits with just "..".

```
> swa <- gsub("\\.\\.: [0-9]+ [0-9]+ \\.\\.\\.\\.", "..", swa)
```

The final substitution removes some unnecessary characters, again to save on space. That is a little complex at this stage but illustrates the power of R for string processing (as well as statistics).

```
> swa <- gsub("( \\$|:|)", "", swa)
```

We use cat() to then display the results of this processing.

```
> cat(swa)

 1 Date           Date, format "2007-11-01" ...
 2 Location       Factor w/ 46 levels "Adelaide","Albany",..
 3 MinTemp        num  8 14 ...
 4 MaxTemp        num  24.3 26.9 ...
 5 Rainfall       num  0 3.6 ...
 6 Evaporation    num  3.4 4.4 ...
 7 Sunshine       num  6.3 9.7 ...
 8 WindGustDir    Ord.factor w/ 16 levels "N"<"NNE"<"NE"<..
 9 WindGustSpeed  num  30 39 ...
10 WindDir9am     Ord.factor w/ 16 levels "N"<"NNE"<"NE"<..
11 WindDir3pm     Ord.factor w/ 16 levels "N"<"NNE"<"NE"<..
12 WindSpeed9am   num  6 4 ...
13 WindSpeed3pm   num  20 17 ...
14 Humidity9am    int  68 80 ...
15 Humidity3pm    int  29 36 ...
16 Pressure9am    num  1020 ...
17 Pressure3pm    num  1015 ...
18 Cloud9am       int  7 5 ...
19 Cloud3pm       int  7 3 ...
20 Temp9am        num  14.4 17.5 ...
21 Temp3pm        num  23.6 25.7 ...
22 RainToday      Factor w/ 2 levels "No","Yes" 1 2 ...
23 RISK_MM        num  3.6 3.6 ...
24 RainTomorrow   Factor w/ 2 levels "No","Yes" 2 2 ...
```

3.8 Summary

In this chapter, we have introduced the concepts of data and dataset. We have described how we obtain data and issues related to the data we use for data mining. We have also introduced some basic data manipulation using R. We will revisit the *weather*, *weatherAUS*, and *audit* datasets throughout the book. Appendix B describes in detail how these datasets are obtained and processed into a form for use in data mining. The amount of detail there and the R code provided may be useful in learning more about manipulating data in R.

3.9   Command Summary

This chapter has referenced the following R packages, commands, functions, and datasets:

<-	function	Assign a value into a named reference.
$	function	Extract a variable from a dataset.
audit	dataset	Sample dataset from **rattle**.
c()	function	Combine items to form a collection.
cat()	function	Display the arguments to the screen.
dim()	function	Report the rows and columns of a dataset.
gsub()	function	Globally substitute one string for another.
head()	function	Show top observations of a dataset.
library()	command	Load a package into the R library.
names()	function	Show variables contained in a dataset.
paste()	function	Combine strings into one string.
rattle	package	Provides the *weather* and *audit* datasets.
seq()	function	Generate a sequence of numbers.
sprintf	function	Format a string with substitution.
str()	function	Show the structure of an object.
weather	dataset	Sample dataset from **rattle**.
weatherAUS	dataset	A larger dataset from **rattle**.

Chapter 4

Loading Data

Data can come in many different formats from many different sources. By using R's extensive capabilities, Rattle provides direct access to such data. Indeed, we are fortunate with the R system in that it is an open system and therefore is strong on sharing and cooperating with other applications. R supports importing data in many formats.

One of the most common formats for data exchange between applications is the comma-separated value (CSV) file. Such files typically have a csv filename extension. This is a simple text file format that is oriented around rows and columns, using a comma to separate the columns in the file. Such files can be used to transfer data through export and import between spreadsheets, databases, weather monitoring stations, and many other applications. A variation on the idea is to separate the columns with other markers, such as a tab character, which is often associated with files having a txt filename extension.

These simple data files (the CSV and TXT files) contain no explicit metadata information—that is, there is no data to describe the structure of the data contained in the file. That information often needs to be guessed at by the software reading the data.

Other types of data sources do provide information about the data so that our software does not need to make guesses about what it is reading. Attribute-Relation File Format files (Section 4.2) have an arff filename extension and add metadata to the CSV format.

Extracting data directly from a database often delivers the metadata along with the data itself. The Open Database Connectivity (ODBC) standard provides an open access method for accessing data stored in a variety of databases and is supported by R. This allows direct connection to a vast collection of data sources, including Microsoft Excel, Microsoft Access, SQL Server, Oracle, MySQL, Postgres, and SQLite. Section 4.3 covers the package **RODBC**.

The full variety of R's capability for loading data is necessarily not available directly within Rattle. However, we can use the underlying R commands to load data and then access it within Rattle, as in Section 4.4.

R packages themselves also provide an extensive collection of sample datasets. Whilst many datasets will be irrelevant to our specific tasks, they can be used to experiment with data mining using R. A list of datasets contained in the R Library is available through the Rattle interface by choosing Library as the Source on the Data tab. We cover this further in Section 4.6.

Having loaded our data into Rattle through some mechanism, we need to decide on the role played by each of the variables in the dataset. We also need to decide how the observations in the dataset are going to be used in the mining. We record these decisions through the Rattle interface, with Rattle itself providing useful defaults.

Once a dataset source has been identified and the Data tab executed, an overview of the data will be displayed in the text view. Figure 4.1 displays the Rattle application after loading the `weather.csv` file, which is supplied as a sample dataset with the Rattle package. We get here by starting up R and then loading **rattle**, starting up Rattle, and then clicking the Execute button for an offer to load the *weather* dataset:

```
> library(rattle)
> rattle()
```

In this chapter, we review the different source data formats and discuss how to load them for data mining. We then review the options that Rattle provides for identifying how the data is to be used for data mining.

4.1 CSV Data

One of the simplest and most common ways of sharing data today is via the comma-separated values (CSV) format. CSV has become a standard file format used to exchange data between many different applications. CSV files, which usually have a `csv` extension, can be exported and imported by spreadsheets and databases, including LibreOffice Calc, Gnumeric, Microsoft Excel, SAS/Enterprise Miner, Teradata, Netezza, and

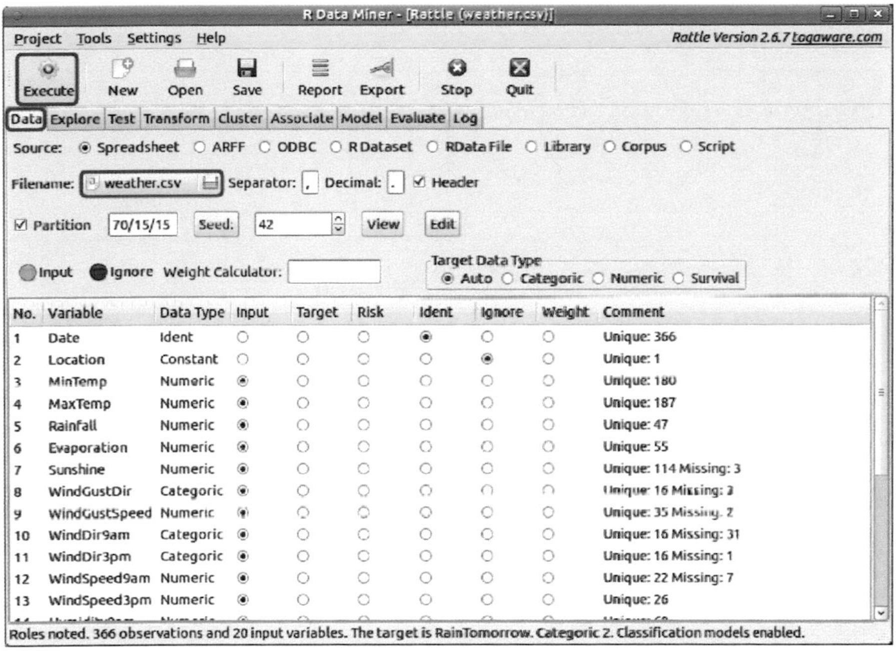

Figure 4.1: Loading the `weather.csv` dataset.

very many other applications. For these reasons, CSV is a good option for importing data into Rattle. The downside is that a CSV file does not contain explicit metadata (i.e., data about the data—including whether the data is numeric or categoric). Without this metadata, R sometimes determines the wrong data type for a particular column. This is not usually fatal, and we can help R along when loading data using R.

Locating and Loading Data

Using the Spreadsheet option of Rattle's Data tab, we can load data directly from a CSV file. Click the Filename button (Figure 4.2) to display the file chooser dialogue (Figure 4.3). We can browse to the CSV file we wish to load, highlight it, and click the Open button.

We now need to actually load the data into Rattle from the file. As always, we do this with a click on the Execute button (or a press of the ⌈2 key). This will load the contents of the file from the hard disk into the computer's memory for processing by Rattle as a dataset.

Rattle supplies a number of sample CSV files and in particular pro-

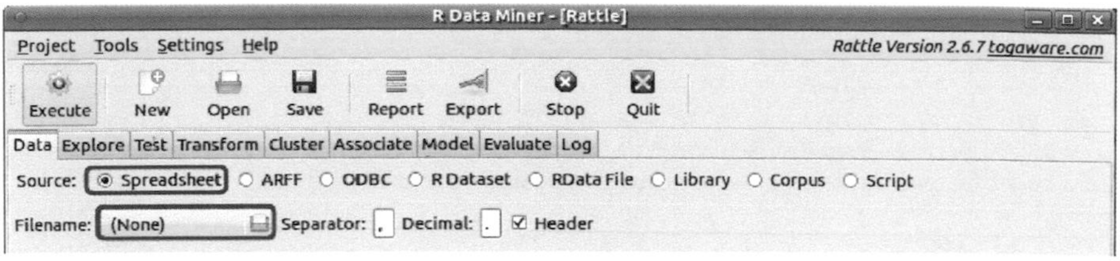

Figure 4.2: The Spreadsheet option of the Data tab, highlighting the Filename button. Click this button to open up the file chooser.

vides the `weather.csv` data file. The data file will have been installed when **rattle** was installed. We can ask R to tell us the actual location of the file using `system.file()`, which we can type into the R Console:

```
> system.file("csv", "weather.csv", package="rattle")
[1] "/usr/local/lib/R/site-library/rattle/csv/weather.csv"
```

The location reported will depend on your particular installation and operating system. Here the location is relative to a standard installation of a Ubuntu GNU/Linux system.

> ***Tip:*** *We can also load this file into a new instance of **Rattle** with just two mouse clicks (**Execute** and **Yes**). We can then click the **Filename** button (displaying* `weather.csv`*) to open up a file browser showing the file path at the top of the window.*

We can review the contents of the file using `file.show()`. This will pop up a window displaying the contents of the file:

```
> fn <- system.file("csv", "weather.csv", package="rattle")
> file.show(fn)
```

The file contents can be directly viewed outside of R and Rattle with any simple text editor. If you aren't familiar with CSV files, it is instructional to become so. We will see that the top of the file begins:

```
Date,Location,MinTemp,MaxTemp,Rainfall,Evaporation...
2007-11-01,Canberra,8,24.3,0,3.4,6.3,NW,30,SW,NW...
2007-11-02,Canberra,14,26.9,3.6,4.4,9.7,ENE,39,E,W...
2007-11-03,Canberra,13.7,23.4,3.6,5.8,3.3,NW,85,N,NNE...
```

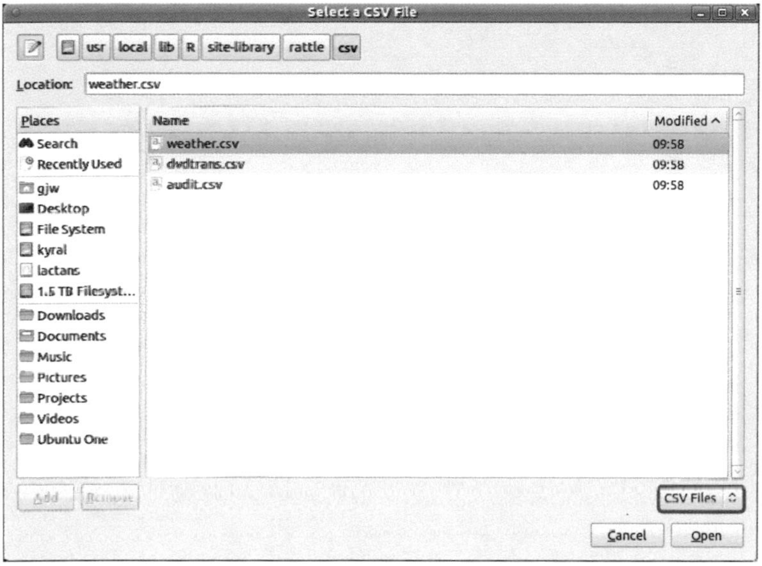

Figure 4.3: The CSV file chooser showing just those files with a .csv extension in the folder. We can also select to display just the .txt files (e.g., the extension often used for tab-delimited files) or else all files by selecting from the drop-down menu at the bottom right.

A CSV file is just a normal text file that commonly begins with a header line listing the names of the variables, each separated by a comma. The remainder of the file after the header row is expected to consist of rows of data that record the observations. For each observation, the fields are separated by commas, delimiting the actual observation of each of the variables.

Loading data into Rattle from a CSV file uses read.csv(). We can see this by reviewing the contents of the Log tab. From the Log tab we will see something like the following:

```
> crs$dataset <- read.csv("file:.../weather.csv",
                          na.strings=c(".", "NA", "", "?"),
                          strip.white=TRUE)
```

The full path to the weather.csv file is truncated here for brevity, so the command above won't succeed with a copy-and-paste. Instead, copy the corresponding line from the Log tab into the R Console. The result of executing this function is that the dataset itself is loaded into memory and referenced using the name `crs\$dataset`.

The second argument in the function call above (`na.strings=`) lists the four strings that, if found as the value of a variable, will be translated into R's representation for missing values (`NA`). The list of strings used here captures the most common approaches to representing missing values. SAS, for example, uses the dot ("."') to denote missing values, and R uses the special string "NA". Other applications simply use the empty string, whilst yet others (including machine learning applications like C4.5) use the question mark ("?").

We also use the `strip.white=` argument, setting it to TRUE, which has the effect of stripping white space (i.e., spaces and/or tabs). This allows the source CSV file to have the commas aligned for easier human viewing and still support missing values appropriately.

The `read.csv()` function need not be quite so complex. If we have a CSV file to load into R (again substituting the "..." with the actual path to the file), we can usually simply type the following command:

```
> ds <- read.csv(".../weather.csv")
```

We can also load data directly from the Internet. For example, the weather dataset is available from `togaware.com`:

```
> ds <- read.csv("http://rattle.togaware.com/weather.csv")
```

As we saw in Chapter 2 Rattle will offer to load the supplied sample data file (`weather.csv`) if no other data file is specified through the Filename button. This is the simplest way to load sample data into Rattle, and is useful for learning the Rattle interface.

After identifying the file to load, we do need to remember to click the Execute button to actually load the dataset into Rattle. The main text panel of the Data tab then changes to list the variables, together with their types and roles and some other useful information, as can be seen in Figure 4.1.

After loading the data from the file into Rattle, thereby creating a dataset, we can begin to explore it. The top of the file can be viewed in the R Console, as we saw in Chapter 2. Here we limit the display to just the first five variables and request just six observations:

```
> head(crs$dataset[1:5], 6)

       Date Location MinTemp MaxTemp Rainfall
1 2007-11-01 Canberra     8.0    24.3      0.0
2 2007-11-02 Canberra    14.0    26.9      3.6
3 2007-11-03 Canberra    13.7    23.4      3.6
4 2007-11-04 Canberra    13.3    15.5     39.8
5 2007-11-05 Canberra     7.6    16.1      2.8
6 2007-11-06 Canberra     6.2    16.9      0.0
```

As we described earlier (Section 2.9, page 50), Rattle stores the dataset within an environment called crs, so we can reference it directly in R as crs$dataset.

Through the Rattle interface, once we have loaded the dataset, we can also view it as a spreadsheet by clicking the View button, which uses dfedit() from **RGtk2Extras** (Taverner et al., 2010).

Data Variations

The Rattle interface provides options for tuning how we read the data from a CSV file. As we can see in Figure 4.2, the options include the Separator and Header.

We can choose the field delimiter through the Separator entry. A comma is the default. To load a TXT file, which uses a tab as the field separator, we replace the comma with the special code \\t (that is, two slashes followed by a t) to represent a tab. We can also leave the entry empty and any white space (i.e., any number of spaces and/or tabs) will be used as the separator.

From the read.csv() viewpoint, the effect of the separator entry is to include the appropriate argument (using sep=) in the call to the function. In this example, if we happen to have a file named "mydata.txt" that contained tab-delimited data, then we would include the sep=:

```
> ds <- read.csv("mydata.txt", sep="\t")
```

> **Tip:** Note that when specifying the tab as the separator directly within R we use a single slash rather than the double slashes through the **Rattle** interface.

Another option of interest when loading a dataset is the Header check box. Generally, a CSV file will have as its first row a list of column names.

These names will be used by R and Rattle as the names of the variables. However, not all CSV files include headers. For such files, uncheck the Header check box. On loading a CSV file that does not contain headers, R will generate variable names for the columns. The check box translates to the header= argument in the call to read.csv(). Setting the value of header= to FALSE will result in the first line being read as data rather than as a header line. If we had such a file, perhaps called "mydata.csv", then the call to read.csv() would be:

```
> ds <- read.csv("mydata.csv", header=FALSE)
```

> *Tip:* *The data can contain different numbers of columns in different rows, with missing columns at the end of a row being filled with NAs. This is handled using the fi l l= argument of read.csv(), which is TRUE by default.*

Basic Data Summary

Once a dataset has been loaded into Rattle, we can start to obtain an idea of the shape of the data from the simple summary that is displayed. In Figure 4.1, for example, the first variable, Date, is recognised as a unique identifier for each observation. It has 366 unique values, which is the same as the number of observations.

The variable Location has only a single value across all observations in the dataset. Consequently, it is identified as a constant and plays no role in the modelling. It is ignored.

The next five variables in Figure 4.1 are all tagged as numeric, followed by the categoric WindGustDir, and so on. The Comment column identifies the unique number of values and the number of missing observations for each variable. Sunshine, for example, has 114 unique values and 3 missing values. How to deal with missing values is covered in Chapter 7.

4.2   ARFF Data

The Attribute-Relation File Format (ARFF) is a text file format that is essentially a CSV file with a number of rows at the top of the file that contain metadata. The ARFF format was developed for use in the

Weka (Witten and Frank, 2005) machine learning software, and there are many datasets available in this format. We can load an ARFF dataset into Rattle through the ARFF option (Figure 4.4), specifying the filename from which the data is loaded.

Rattle provides sample ARFF datasets. To access them, after starting up Rattle and loading the sample *weather* dataset (Section 2.4), choose the ARFF option and then click the Filename chooser. Browse to the parent folder and then into the `arff` folder to choose a dataset to load.

Figure 4.4: Choosing the ARFF radio button to load an ARFF file.

The key difference between CSV and ARFF is in the top part of the file, which contains information about each of the variables in the data— this is the data description section. An example of the ARFF format for our *weather* dataset is shown below. Note that ARFF refers to variables as *attributes*.

```
@relation weather
@attribute Date date
@attribute Location {Adelaide, Albany, ...}
@attribute MinTemp numeric
@attribute MaxTemp numeric
...
@attribute RainTomrrow {No, Yes}
@data
2010-11-01,Canberra,8,24.3,0,...,Yes
2010-11-02,Canberra,14,26.9,3.6,...,Yes
2010-11-03,Canberra,?,23.4,3.6,...,Yes
...
```

The data description section is straightforward, beginning with the name of the dataset (or the name of the *relation* in ARFF terminology). Each of the variables used to describe each observation is then identified together with its data type. Each variable definition appears on a single line (we have truncated the lines in the example above). Numeric variables are identified as numeric, real, or integer. For categoric variables, the possible values are listed.

Two other data types recognised by ARFF are string and date. A string data type indicates that the variable can have a string (a sequence of characters) as its value. The date data type can also optionally specify the format in which the date is recorded. The default for dates is the ISO-8601 standard format, which is "yyyy-MM-dd'T'HH:mm:ss".

Following the metadata specification, the actual observations are then listed, each on a single line, with fields separated by commas, exactly as with a CSV file.

A significant advantage of the ARFF data file over the CSV data file is the metadata information. This is particularly useful in Rattle, where for categoric data the possible values are determined from the data when reading in a CSV file. Any possible values of a categoric variable that are not present in the data will, of course, not be known. When reading the data from an ARFF file, the metadata will list all possible values of a categoric variable, even if one of the values might not be used in the actual data. We will come across this as an issue, particularly when we build and deploy random forest models, as covered in Chapter 12.

Comments can also be included in an ARFF file with a "%" at the beginning of the comment line. Including comments in the data file allows us to record extra information about the dataset, including how it was derived, where it came from, and how it might be cited.

Missing values in an ARFF data file are identified using the question mark "?". These are identified by R's read.arff(), and we see them as the usual NAs in Rattle.

Overall, the ARFF format, whilst simple, is quite an advance over a CSV file. Nonetheless, CSV still remains the more common data file.

4.3   ODBC Sourced Data

Much data is stored within databases and data warehouses. The Open Database Connectivity (ODBC) standard has been developed as a common approach for accessing data from databases (and hence data warehouses). The technology is based on the Structured Query Language (SQL) used to query relational databases. We discuss here how to access data directly from such databases.

Rattle can obtain a dataset from any database accessible through ODBC by using Rattle's ODBC option (Figure 4.5). Underneath the GUI, **RODBC** (Ripley and Lapsley, 2010) provides the actual interface to the ODBC data source.

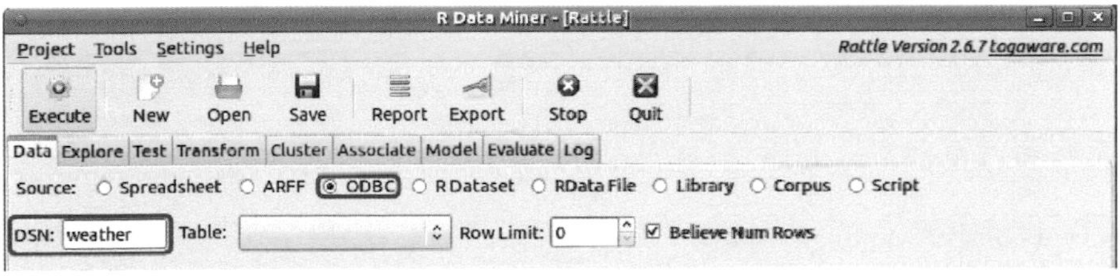

Figure 4.5: Loading data through an ODBC database connection.

The key to accessing data via ODBC is to identify the data source through a so-called data source name (or DSN). Different operating systems provide different mechanisms for setting up DSNs. Under the GNU/Linux operating system, for example, using the unixodbc application, the system DSNs are often defined in /etc/odbcinst.ini and /etc/odbc.ini. Under Microsoft Windows, the control panel provides access to the ODBC Data Sources tool.

Using Rattle, we identify a configured DSN by typing its name into the DSN text entry (Figure 4.5). Once a DSN is specified, Rattle will attempt to make a connection. Many ODBC drivers will prompt for a username and password before establishing the connection. Figure 4.6 illustrates a typical popup for entering such data, in this case for connecting to a Netezza data warehouse.

To establish a connection using R directly, we use odbcConnect() from **RODBC**. This function establishes what we might think of as a channel connecting to the remote data source:

```
> library(RODBC)
> channel <- odbcConnect("myDWH", uid="kayon", pwd="toga")
```

After establishing a connection to a data source, Rattle will query the database for the names of the available tables and provide access to that list through the Table combo box of Figure 4.5. We need to select the specific table to load.

A limited number of options available in R are exposed through Rattle for fine-tuning the ODBC connection. One option allows us to limit the

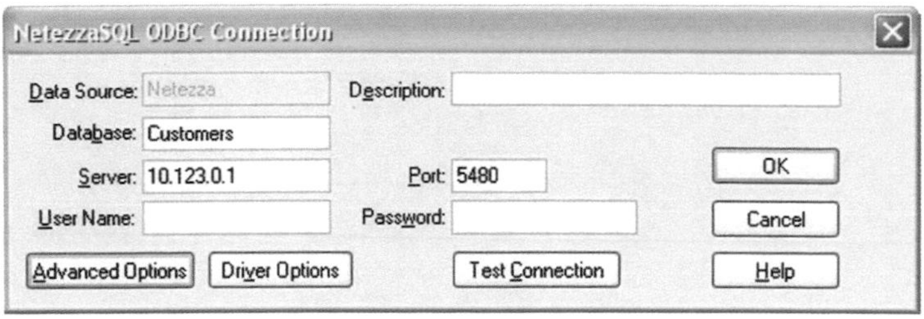

Figure 4.6: Netezza ODBC connection

number of rows retrieved from the chosen table. If the row limit is set
to 0, then all of the rows from the table are retrieved. Unfortunately,
there is no SQL standard for limiting the number of rows returned from
a query. For some database systems (e.g., Teradata and Netezza), the
SQL keyword is LIMIT, and this is what is used by Rattle.

A variety of R functions, provided by **RODBC**, are available to interact with the database. For example, the list of available tables is
obtained using sqlTables(). We pass to it the channel that we created
above to communicate with the database:

```
> tables <- sqlTables(channel)
```

If there is a table in the connected database called, for example,
clients, we can obtain a list of column names using sqlColumns():

```
> columns <- sqlColumns(channel, "clients")
```

Often, we are interested in loading only a specific subset of a table from
the database. We can directly formulate an SQL query to retrieve just
the data we want. For example:

```
> query <- "SELECT * FROM clients WHERE cost > 2500"
> myds <- sqlQuery(channel, query)
```

Using R directly provides a lot more scope for carefully identifying the
data we wish to load. Any SQL query can be substituted for the simple SELECT statement used above. For those with skills in writing SQL
queries, this provides quite a powerful mechanism for refining the data
to be loaded, before it is loaded.

Loading data by directly sending an SQL query to the channel as
above will store the data in R as a dataset, which we can reference as

myds (as defined above). This dataset can be accessed in Rattle with the R Dataset option, which we now introduce.

4.4 R Dataset—Other Data Sources

Data can be loaded from any source, one way or another, into R. We have covered loading data from a data file (as in loading a CSV or TXT file) or directly from a database. However, R supports many more options for importing data from a variety of sources.

Rattle can use any dataset (technically, any data frame) that has been loaded into R as a dataset to be mined. When choosing the R Dataset option of the Data tab (Figure 4.7), the Data Name box will list each of the available data frames that can be brought into Rattle as a dataset.

Using **foreign** (DebRoy and Bivand, 2011), for example, R can be used to read SPSS datasets (read.spss()), SAS XPORT format datasets (read.xport()), and DBF database files (read.dbf()). One notable exception, though, is the proprietary SAS dataset format, which cannot be loaded unless we have a licensed copy of SAS to read the data for us.

Loading SPSS Datasets

As an example, suppose we have an SPSS data file saved or exported from SPSS. We can read that into R using read.spss():

```
> library(foreign)
> mydataset <- read.spss(file="mydataset.sav")
```

Then, as in Figure 4.7, we can find the data frame name, mydataset, listed as an available R Dataset:

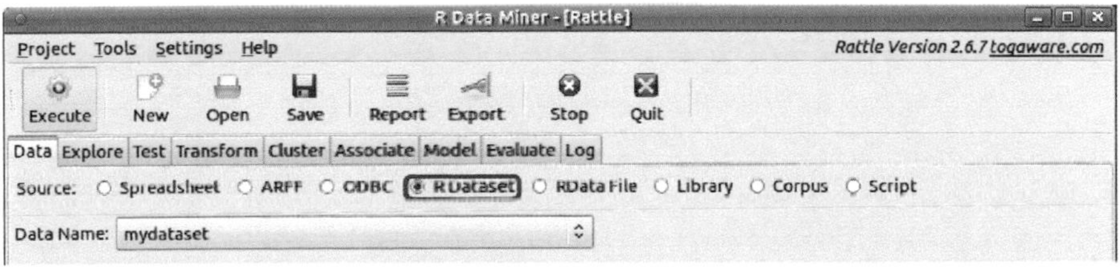

Figure 4.7: Loading an already defined R data frame as a dataset for use in Rattle.

The datasets that we wish to use with Rattle need to be constructed or loaded into the same R session that is running Rattle (i.e., the same R Console in which we loaded the Rattle package).

Reading Data from the Clipboard

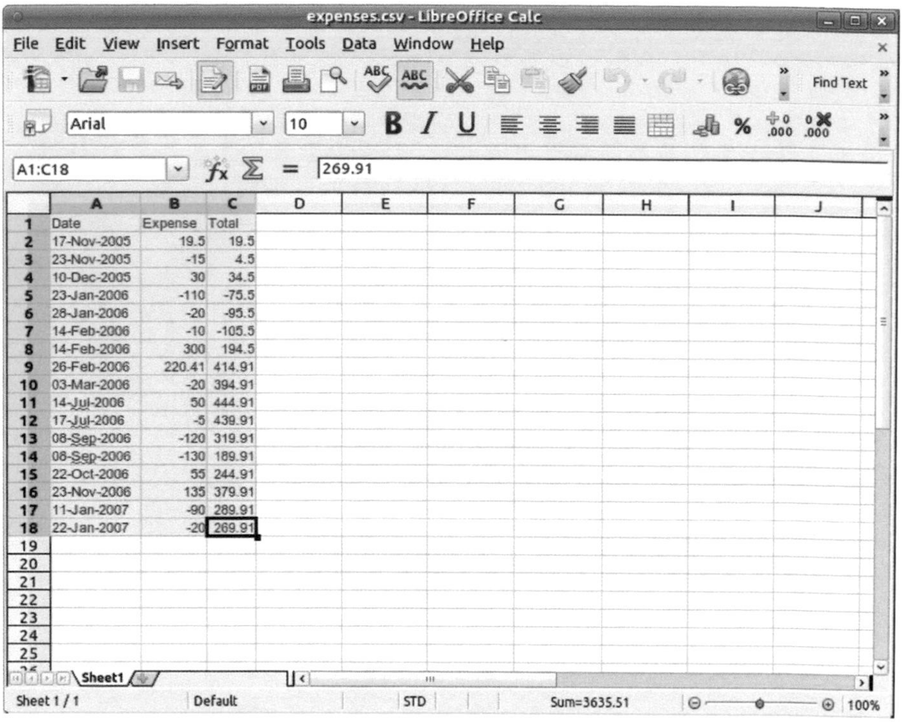

Figure 4.8: Selected region of a spreadsheet copied to the clipboard.

An interesting variation that may at times be quite convenient is the ability to directly copy and paste a selection via the system clipboard. Through this mechanism, we can "copy" (as in "copy-and-paste") data from a spreadsheet into the clipboard. Then, within R we can "paste" the data into a dataset using read.table().

Suppose we have opened a spreadsheet with the data we see in Figure 4.8. If we select the 16 rows, including the header, in the usual way, we can very simply load the data using R:

```
> expenses <- read.table(file("clipboard"), header=TRUE)
```

The **expenses** data frame is then available to Rattle.

Converting Dates

By default, the Date variable in the example above is loaded as categoric. We can convert it into a date type, as below, before we load it into Rattle, as in Figure 4.9:

```
> expenses$Date <- as.Date(expenses$Date,
                           format="%d-%b-%Y")
> head(expenses)

        Date Expense   Total
1 2005-11-17    19.5    19.5
2 2005-11-23   -15.0     4.5
3 2005-12-10    30.0    34.5
4 2006-01-23  -110.0   -75.5
5 2006-01-28   -20.0   -95.5
6 2006-02-14   -10.0  -105.5
```

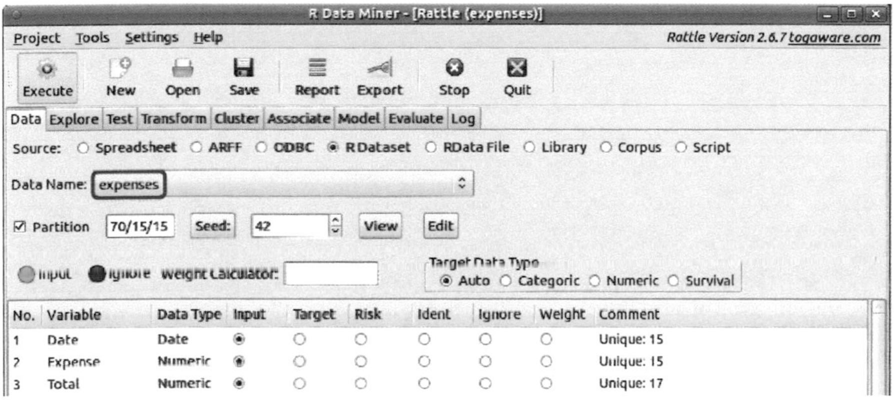

Figure 4.9: Loading an R data frame that was obtained from a copy-and-paste, via the clipboard, from a spreadsheet.

Reading Data from the World Wide Web

A lot of data today is available in HTML format on the World Wide Web. **XML** (Lang, 2011) provides functions to read such data directly into R and so make that data available for analysis in Rattle (and, of course,

R). As an example, we can read data from Google's list of most visited web sites, converting it to a data frame and thus making it available to Rattle. We begin this by loading **XML** and setting up some locations:

```
> library(XML)
> google <- "http://www.google.com/"
> path <- "adplanner/static/top1000/"
> top1000urls <- paste(google, path, sep="")
```

Now we can read in the data using readHTMLTable(), extracting the relevant table and setting up the column names:

```
> tables <- readHTMLTable(top1000urls)
> top1000 <- tables[[2]]
> colnames(top1000) <- c('Rank', 'Site', 'Category',
                         'Users', 'Reach', 'Views',
                         'Advertising')
```

The top few rows of data from the table can be viewed using head():

```
> head(top1000)

  Rank           Site                      Category
1    1  facebook.com           Social Networks
2    2   youtube.com                Online Video
3    3     yahoo.com                 Web Portals
4    4      live.com             Search Engines
5    5 wikipedia.org Dictionaries & Encyclopedias
6    6       msn.com                 Web Portals
        Users Reach             Views Advertising
1 880,000,000 47.2% 910,000,000,000         Yes
2 800,000,000 42.7% 100,000,000,000         Yes
3 660,000,000 35.3%  77,000,000,000         Yes
4 550,000,000 29.3%  36,000,000,000         Yes
5 490,000,000 26.2%   7,000,000,000          No
6 450,000,000   24%  15,000,000,000         Yes
```

4.5   R Data

Using the RData File option (Figure 4.10), data can be loaded directly from a native binary R data file (usually with the RData filename extension). Such files may contain multiple datasets (usually in a compressed format) and will have been saved from R sometime previously (using save()).

RData can be loaded by first identifying the file containing the data. The data will be loaded once the file is identified, and we will be given an option to choose just one of the available data frames to be loaded as Rattle's dataset. We specify this through the Data Name combo box and then click Execute to make the dataset available within Rattle.

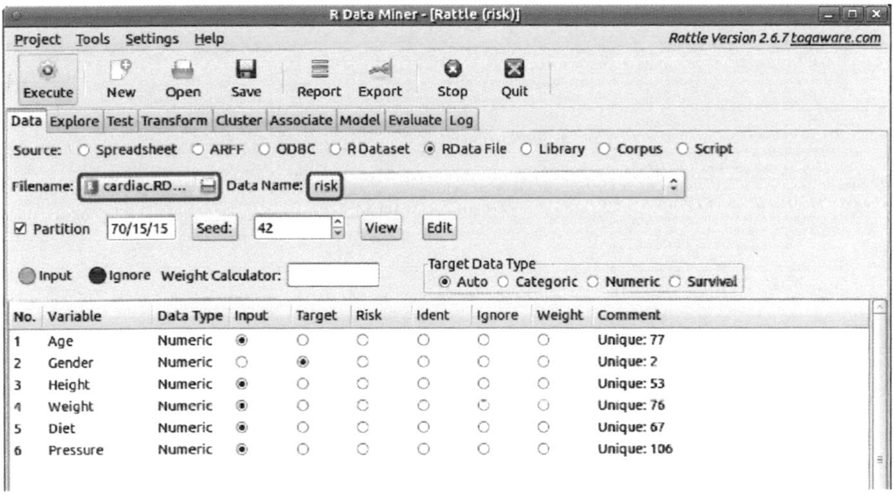

Figure 4.10: Loading a dataset from a binary R data file.

Figure 4.10 illustrates the selection of an RData file. The file is called cardiac.RData. Having identified the file, Rattle will populate the Data Name combo box with the names of each of the data frames found in the file. We can choose the *risk* dataset, from within the data file, to be loaded into Rattle.

4.6 Library

Almost every R package provides a sample dataset that is used to illustrate the functionality of the package. Rattle, as we have seen, provides the *weather*, *weatherAUS*, and *audit* datasets. We can explore the wealth of datasets that are available to us through the packages that are contained in our installed R library.

The Library option of the Data tab provides access to this vast collection of sample datasets. Clicking the radio button will generate the list of available datasets, which can then be accessed from the Data Name combo box. The dataset name, the package that provides that dataset, and a short description of the dataset will be included in the list. Note that the list can be quite long, and its contents will depend on the packages that are installed. We can see a sample of the list here, illustrating the R code that Rattle uses to generate the list:

```
> da <- data(package=.packages(all.available=TRUE))
> sort(paste(da$results[, "Item"], " : ",
             da$results[, "Package"], " : ",
             da$results[, "Title"], sep=""))

...
[10] "Adult : arules : Adult Data Set"
...
[12] "Affairs : AER : Fair's Extramarital Affairs Data"
...
[14] "Aids2 : MASS : Australian AIDS Survival Data"
...
[19] "airmay : robustbase : Air Quality Data"
...
[23] "ais : DAAG : Australian athletes data set"
...
[66] "audit : rattle : Sample dataset for data mining"
...
[74] "Baseball : vcd : Baseball Data"
...
[1082] "weather : rattle : Sample dataset for ..."
...
```

To access a dataset provided by a particular package, the actual package will first need to be loaded using library() (Rattle will do so automatically). For many packages (specifically those that declare the datasets as being *lazy loaded*—that is, loaded when they are referenced), the dataset will then be available from the R Console simply by typing the dataset name. Otherwise, data() needs to be run before the dataset can be accessed. We need to provide data() with the name of the dataset to be made available. Rattle takes care of this for us to ensure the appropriate action is taken to have the dataset available.

4.7 Data Options

All of Rattle's data load options that we have described above share a common set of further options that relate to the dataset once it has been loaded. The additional options relate to sampling the data as well as deciding on the role played by each of the variables. We review these options in the context of data mining.

Partitioning Data

As we first saw in Section 2.7, the Partition option allows us to partition our dataset into a training dataset, a validation dataset, and a testing dataset. The concept of partitioning a dataset was further defined in Section 3.1. The concepts are primarily oriented towards predictive data mining. Generally we will build a model using the *training dataset*.

To evaluate (Chapter 15) the performance of the model, we might then apply it to the *evaluation dataset*. This dataset has not been used to build the model and so provides an estimate of how well the model will perform when presented with new observations. Depending on the performance, we may tune the model-building parameters to seek an improvement in model performance.

Once we have a model that appears to perform well, or as well as possible with respect to the validation dataset, we might then evaluate its performance on the third partition, the *testing dataset*. The model has not previously been exposed to the observations contained in the testing dataset. Thus, the performance of the model on this dataset is probably a very good indication of how well the model will perform on new observations as they become available.

The concept of partitioning or sampling, though, is more general than simply a mechanism for partitioning for predictive data mining purposes. Statisticians have developed an understanding of sampling as a mechanism for analysing a small dataset to make conclusions about the whole population. Thus there is much literature from the statistics community on ensuring a good understanding of the uncertainty surrounding any conclusions we might make from analyses performed on any data. Such an understanding is important, though often underplayed in the data mining context.

Rattle creates a random partition/sample using `sample()`. A random sample will generally have a good chance of reflecting the distributions of the whole population. Thus, exploring the data, as we do in Chapter 5, will be made easier when very large datasets are sampled down into much smaller ones. Exploring 10,000 observations is often a more interactive and practical proposition than exploring 1,000,000 observations. Other advantages of sampling include allowing analyses or plots to be repeated over different samples to gauge the stability and statistical accuracy of results. Model building, as we will see particularly when building random forests (Chapter 12), can take advantage of this.

The use of sampling in this way will also be necessary in data mining when the datasets available to model are so large that model building may take a considerable amount of time (hours or days). Sampling down to small proportions of a dataset will allow us to experiment more inter-actively with building a model. Once we are sure of how the data needs to be cleaned and transformed from our initial interactions, we can start experimenting with models. After we have the basic model parameters in place, we might be in a position to clean, transform, and model over a much larger portion of the data. We can leave the model building to complete over the possibly many hours that are often needed.

The downside of sampling, particularly in the data mining context, is that observations that correspond to rare events might disappear from a sample. Cases of rare diseases, or of the few instances of fraud from amongst millions of electronic funds transfers, may well be lost, even though they are the items that are of most interest to us in many data mining projects. This problem is often referred to as the class imbalance problem.

Rattle provides a default random partitioning of a dataset with 70% of the data going into a training dataset, 15% into a validation dataset, and 15% into a testing dataset (see Figure 4.11). We can override these choices, depending on our needs. A very small sampling may be required to perform some explorations of otherwise very large datasets. Smaller samples may also be required to build models using some of the more computationally expensive algorithms (like support vector machines).

Random numbers are used to select samples. Any sequence of ran-dom numbers must start with a so-called seed. If we use the same seed each time we will get the same sequence of random numbers. Thus the process is repeatable. By changing the seed we can select different ran-dom samples. This is often useful when we wish to explore the sensitivity of our models to different data samples.

Within Rattle a default seed is always used. This ensures, for example,

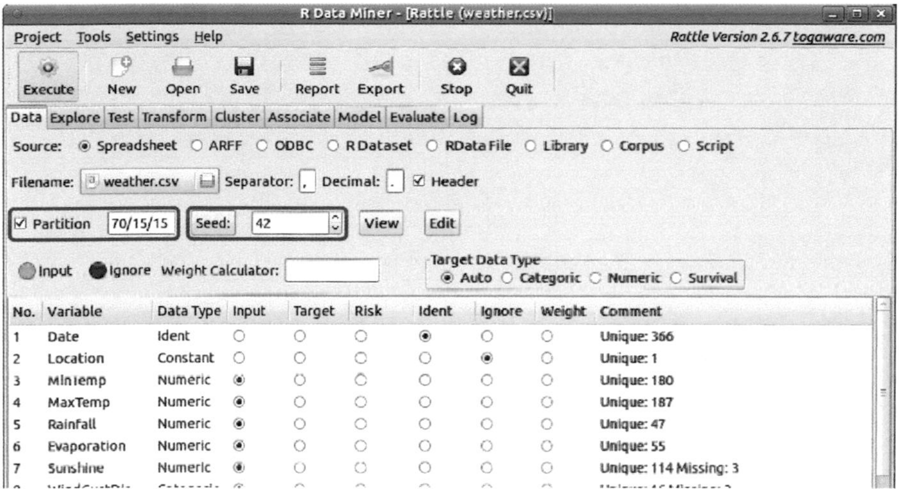

Figure 4.11: Sampling the *weather* dataset.

repeatable modelling. The seed is passed to the R function `set.seed()` to set a seed for the next generated sequence of random numbers. Thus, by setting the seed to the same number each time we can be assured of obtaining the same sample.

Conversely, we may like to set the seed to a different number in a series of model building exercises, and to then compare the performance of each model. Each model will have been built from a different random sample of the dataset. If we see significant variation between the different models, we may be concerned about the robustness of the approach we are taking. We discuss this further in Chapter 15.

Variable Roles

When building a model each variable will play a specific role. Most variables will be inputs to the model, and one variable is often identified as the target which we are modelling.

A variable can also be identified as a so-called *risk variable*. A risk variable might not be used for modelling as such. Generally it will record some magnitude associated with the risk or outcome. In the *audit* dataset, for example, it records the dollar amount of an adjustment that results from an audit—this is a measure of the size of the risk associated with that case. In the *weather* dataset the risk variable is the amount of rain recorded for the following day—the amount of rain can be thought of as the size of the risk. See Section 15.4 for an example of using risk variables within Rattle, specifically for model evaluation.

Finally, we might also identify some variables to be ignored in the modelling altogether.

In loading data into Rattle we need to ensure our variables have their correct role for modelling. The default role for most variables is that of an Input variable. Generally, these are the variables that will be used to predict the value of a Target variable.

A *target variable*, if there is one associated with the dataset, is generally the variable of interest, from a predictive modelling point of view. That is, it is a variable that records the outcome from the historic data. In the case of the *weather* dataset this is RainTomrrow, whilst for the *audit* dataset the target is Adjusted.

Rattle uses simple heuristics to guess at a variable having a Target role. The primary heuristic is that a variable with a small number of distinct values (e.g., less than 5) is considered as a candidate target variable. The last variable in the dataset is usually considered as a candidate for being the target variable, because in many public datasets the last variable often is the target variable. If it has more than 5 distinct values Rattle will proceed from the first variable until it finds one with less than 5, if there are any. Only one variable can be tagged as a Target.

In a similar vain, integer variables that have a unique value for each observation are often automatically identified as an Ident (an identifier). Any number of variables can be tagged as being an Ident. All Ident variables are ignored when modelling, but are used after scoring a dataset, when it is being written to a score file, so that the observations that are scored can be identified.

Not all variables in our dataset might be wanted for the particular modelling task at hand. Such variables can be ignored, using the Ignore radio button.

When loading data into Rattle certain special strings are used to identify variable roles. For example, if the variable name starts with ID then the variable is automatically marked as having a role as an Ident. The user can override this.

Similarly, a variable with a name beginning with IGNORE will have the default role of Ignore. And so with RISK and TARGET.

At any one time a target is either treated as categoric or numeric. For a numeric variable chosen as the target, if it has 10 or fewer unique values then Rattle will automatically treat it as a categoric variable (by default). For modelling purposes, the consequence is that only classification type predictive models will be available. To build regression type predictive models we need to override the heuristic by selecting the Numeric radio button of the Data tab.

Weights Calculator and Role

The final data configuration option of the Data tab is the Weight Calculator and the associated Weight role. A single variable can be identified as representing some weight associated with each observation. The Weight Calculator allows us to provide a formula that could involve multiple variables as well as some scaling to give a weight for each observation. For example, with the audit dataset, we might enter a formula that uses the adjustment amount, and this will give more weight to those observations with a larger adjustment.

4.8   Command Summary

This chapter has referenced the following R packages, commands, functions, and datasets:

archetypes	package	Archetypal analysis.
audit	dataset	Sample dataset from **rattle**.
clients	dataset	A fictitious dataset.
data()	command	Make a dataset available to R.
dfedit()	command	Edit a data frame in a spreadsheet.
file.show()	command	Display a data frame.
foreign	package	Access multiple data formats.
library()	command	Load a package into the R library.
file.show()	command	Display the contents of a file.
odbcConnect()	function	Connect to a database.
paste()	function	Combine strings into one string.
rattle	package	Provides sample datasets.

`read.arff()`	function	Read an ARFF data file.
`read.csv()`	function	Read a comma-separated data file.
`read.dbf()`	function	Read data from a DBF database file.
`read.delim()`	function	Read a tab-delimited data file.
`read.spss()`	function	Read data from an SPSS data file.
`read.table()`	function	Read data from a text file.
`read.xport()`	function	Read data from a SAS Export data file.
`readHTMLTable()`	function	Read data from the World Wide Web.
risk	dataset	A fictitious dataset.
RODBC	package	Provides database connectivity.
`sample()`	function	Take a random sample of a dataset.
`save()`	command	Save R objects to a binary file.
`set.seed()`	command	Reset the random number sequence.
skel	dataset	Dataset from **archetypes** package.
`sqlColumns()`	function	List columns of a database table.
`sqlTables()`	function	List tables available from a database.
`system.file()`	function	Locate R or package file.
weather	dataset	Sample dataset from **rattle**.
XML	package	Access and generate XML like HTML.

Chapter 5

Exploring Data

As a data miner, we need to live and breathe our data. Even before we start building our data mining models, we can gain significant insights through exploring the data. Insights gained can deliver new discoveries to our clients—discoveries that can offer benefits early on in a data mining project. Through such insights and discoveries, we will increase our knowledge and understanding.

Through exploring our data, we can discover what the data looks like, its boundaries (the minimum and maximum values), its numeric characteristics (the average value), and how it is distributed (how spread out the data is). The data begins to tell us a story, and we need to build and understand that story for ourselves. By capturing that story, we can communicate it back to our clients.

This task of exploratory data analysis (often abbreviated as EDA) is a core activity in any data mining project. Exploring the data generally involves getting a basic understanding of a dataset through numerous variable summaries and visual plots.

Through data exploration, we begin to understand the "lay of the land" just as a gold miner works to understand the terrain rather than blindly digging for gold randomly. Through this exploration, we will often identify problems with the data, including missing values, noise, erroneous data, and skewed distributions. This in turn will drive our choice of the most appropriate and, importantly, applicable tools for preparing and transforming our data and for mining. Some tools, for example, are limited in use when there is much missing data.

Rattle provides tools ranging from textual summaries to visually appealing graphical plots for identifying correlations between variables. The

Explore tab within Rattle provides a window into the tools for helping us understand our data, driving the many options available in R.

5.1 Summarising Data

Figure 5.1 shows the options available under Rattle's Explore tab. We begin our exploration of data with the basic Summary option, which provides a textual overview of the data. Whilst a picture may be worth a thousand words, textual summaries still play an important role in our understanding of data.

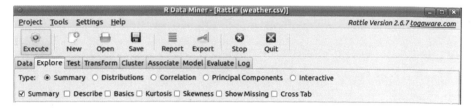

Figure 5.1: The Explore tab provides access to a variety of ways in which we start to understand our data.

Often, we deal with very large datasets, and some of the calculations and visualisations we perform will be computationally quite expensive. Thus it may be useful to summarise random subsets of the data instead. The Partition option of the Data tab is useful here. This uses sample() to generate a list of row numbers that we can then use to index the dataset.

The following example generates a 20% (i.e., 0.2 times the number of rows) random sample of our *weather* dataset. We use nrow() to obtain the number of rows in the sample (73.2) and dim() for information about the number of rows and columns in the data frame:

```
> library(rattle)
> dim(weather)

[1] 366   24

> set.seed(42)
> smpl <- sample(nrow(weather), 0.2*nrow(weather))
> dim(weather[smpl,])

[1] 73 24
```

For our *weather* dataset with only 366 observations, we clearly do not need to sample.

5.1.1   Basic Summaries

The simplest text-based statistical summary of a dataset is provided by summary(). This is always a useful starting point in reviewing our data. It provides a summary of each variable. Here we see summaries for a mixture of numeric and categoric variables:

```
> summary(weather[7:9])

    Sunshine        WindGustDir    WindGustSpeed
Min.    : 0.00   NW       : 73   Min.    :13.0
1st Qu.: 5.95   NNW      : 44   1st Qu.:31.0
Median : 8.60   E        : 37   Median :39.0
Mean    : 7.91  WNW      : 35   Mean    :39.8
3rd Qu.:10.50   ENE      : 30   3rd Qu.:46.0
Max.    :13.60  (Other):144   Max.    :98.0
NA's    : 3.00  NA's     :  3   NA's    : 2.0
```

For the numeric variables, summary() will list the minimum and maximum values together with average values (the mean and median) and the first and third quartiles. The quartiles represent a partitioning of the values of the numeric variable into four equally sized sets. The first quartile includes 25% of the observations of this variable that have a value less than this first quartile. The third quartile is the same, but at the 75% mark. The median is actually also the second quartile, representing the 50% cutoff (i.e., the middle value).

Generally, if the mean and median are significantly different, then we would think that there are some observations of this variable that are quite a distance from the mean in one particular direction (i.e., some exceptionally large positive or negative values, generally called outliers, which we cover in Chapter 7). From the variables we see above, Sunshine has a relatively larger (although still small) gap between its mean and median, whilst the mean and median of WindGustSpeed are quite similar. Sunshine has more small observations than large observations, using our terms rather loosely.

The categoric variables will have listed for them the top few most frequent levels with their frequency counts and then aggregate the remainder under the (Other) label. Thus there are 73 observations with a NW wind gust, 44 with a NNW wind gust, and so on. We observe quite a predominance of these northwesterly wind gusts. For both types of listings, the count of any missing values (NAs) will be reported. A somewhat more detailed summary is obtained from describe(), provided by **Hmisc** (Harrell, 2010). To illustrate this we first load **Hmisc** into the library:

```
> library(Hmisc)
```

For numeric variables like Sunshine (which is variable number 7) describe() outputs two more deciles (10% and 90%) as well as two other percentiles (5% and 95%). The output continues with a list of the lowest few and highest few observations of the variable. The extra information is quite useful in building up our picture of the data.

```
> describe(weather[7])
weather[7]

 1  Variables      366  Observations
---------------------------------------------------------------
Sunshine
       n missing  unique    Mean    .05     .10     .25
     363       3     114   7.909   0.60    2.04    5.95

     .50     .75     .90     .95
    8.60   10.50   11.80   12.60

lowest :  0.0  0.1  0.2  0.3  0.4
highest: 13.1 13.2 13.3 13.5 13.6
---------------------------------------------------------------
```

For categoric variables like WindGustDir (which is variable number 8) describe() outputs the frequency count and the percentage this represents for each level. The information is split over as many lines as is required, as we see in the following code box.

```
> describe(weather[8])

weather[8]

 1  Variables      366  Observations
---------------------------------------------------------------
WindGustDir
      n missing  unique
    363       3      16

           N NNE NE ENE   E ESE SE SSE   S SSW SW WSW   W
Frequency 21   8 16  30  37  23 12  12  22   5  3   2  20
%             6   2  4   8  10   6  3   3   6   1  1   1   6
          WNW NW NNW
Frequency  35 73  44
%          10 20  12
---------------------------------------------------------------
```

5.1.2   Detailed Numeric Summaries

An even more detailed summary of the numeric data is provided by
basicStats() from **fBasics** (Wuertz et al., 2010). Though intended for
time series data, it provides useful statistics in general, as we see in the
code box below.

Some of the same data that we have already seen is presented together
with a little more. Here we see that the variable Sunshine is observed 366
times, of which 3 are missing (NAs). The minimum, maximum, quartiles,
mean, and median are as before.

The statistics then go on to include the total sum of the amount of
sunshine, the standard error of the mean, the lower and upper confidence
limits on the true value of the mean (at a 95% level of confidence), the
variance and standard deviation, and two measures of the shape of the
distribution of the data: skewness and kurtosis (explained below).

The mean is stated as being 7.91. We can be 95% confident that the
actual mean (the *true mean*) of the population, of which the data we
have here is assumed to be a random sample, is somewhere between 7.55
and 8.27.

```
> library(fBasics)
> basicStats(weather$Sunshine)
                X..weather.Sunshine
nobs                    366.0000
NAs                       3.0000
Minimum                   0.0000
Maximum                  13.6000
1. Quartile               5.9500
3. Quartile              10.5000
Mean                      7.9094
Median                    8.6000
Sum                    2871.1000
SE Mean                   0.1827
LCL Mean                  7.5500
UCL Mean                  8.2687
Variance                 12.1210
Stdev                     3.4815
Skewness                 -0.7235
Kurtosis                 -0.2706
```

The standard deviation is a measure of how spread out (or how dispersed or how variable) the data is with respect to the mean. It is measured in the same units as the mean itself. We can read it to say that most observations (about 68% of them) are no more than this distance from the mean. That is, most days have 7.91 hours of sunshine, plus or minus 3.48 hours.

Our observation of the mean and standard deviation for the sunshine data needs to be understood in the context of other knowledge we glean about the variable. Consider again Figure 2.8 on page 34. An observation we might make there is that the distribution appears to be what we might call bimodal—that is it has two distinct scenarios. One is that of a cloudy day, and for such days the hours of sunshine will be quite small. The other is that of a sunny day, for which the hours of sunshine will cover the whole day. This observation might be more important to us in weather forecasting, than the interval around the mean. We might want to transform this variable into a binary variable to capture this observation. Transformations are covered in Chapter 7. The variance is the square of the standard deviation.

5.1.3   Distribution

In statistics, we often talk about how observations (i.e., the values of a variable) are distributed. By "distributed" we mean how many times each value of some variable might appear in a collection of data. For the variable Sunshine, for example, the distribution is concerned with how many days have 8 hours of sunshine, how many have 8.1 hours, and so on.

The concept is not quite that black and white, though. In fact, the distribution is often visualised as a smooth curve, as we might be familiar with from published articles that talk about a *normal* (or some other common) distribution. We often hear about the bell curve. This is a graph that plots a shape similar to that of musical bells. For our discussion here, it is useful to have a mental picture of such a bell curve, where the horizontal axis represents the possible values of the variable (the observations) and the vertical axis represents how often those values might occur.

5.1.4   Skewness

The skewness is a measure of how asymmetrically our data is distributed. The skewness indicates whether there is a long tail on one or the other side of the mean value of the data. Here we use skewness() from **Hmisc** to compare the distributions of a number of variables:

```
> skewness(weather[,c(7,9,12,13)], na.rm=TRUE)

    Sunshine WindGustSpeed  WindSpeed9am  WindSpeed3pm
     -0.7235        0.8361        1.3602        0.5913
```

A skewness of magnitude (i.e., ignoring whether it is positive or negative) greater than 1 represents quite an obvious extended spread of the data in one direction or the other. The direction of the spread is indicated by the sign of the skewness. A positive skewness indicates that the spread is more to the right side of the mean (i.e., above the mean) and is referred to as having a longer right tail. A negative skewness is the same but on the left side.

Many models and statistical tests are based on the assumption of a so-called bell curve distribution of the data, which describes a symmetric spread of data values around the mean. The greater the skewness, the greater the distortion to this spread of values. For a large skewness, the assumptions of the models and statistical tests will not hold, and so we need to be a little more careful in their use. The impact tends to be greater for traditional statistical approaches and less so for more recent approaches like decision trees.

5.1.5   Kurtosis

A companion for skewness is kurtosis, which is a measure of the nature of the peaks in the distribution of the data. Once again, we might picture the distribution of the data as having a shape that is something like that of a church bell (i.e., a bell curve). The kurtosis tells us how skinny or fat the bell is. **Hmisc** provides `kurtosis()`:

```
> kurtosis(weather[,c(7,9,12,13)], na.rm=TRUE)

   Sunshine WindGustSpeed  WindSpeed9am  WindSpeed3pm
    -0.2706        1.4761        1.4758        0.1963
```

A larger value for the kurtosis indicates that the distribution has a sharper peak, primarily because there are only a few values with more extreme values compared with the mean value. Thus, `WindSpeed9am` has a sharper peak and a smaller number of more extreme values than `WindSpeed3pm`. The lower kurtosis value indicates a flatter peak.

5.1.6   Missing Values

Missing values present challenges to data mining and modelling in general. There can be many reasons for missing values, including the fact that the data is hard to collect and so not always available (e.g., results of an expensive medical test), or that it is simply not recorded because it is in fact 0 (e.g., spouse income for a spouse who stays home to manage the family). Knowing why the data is missing is important in deciding how to deal with the missing value.

We can explore the nature of the missing data using `md.pattern()` from **mice** (van Buuren and Groothuis-Oudshoorn, 2011), as Rattle does when activating the Show Missing check button of the Summary option of the Explore tab. The results can help us understand any structure in the missing data and even why the data is missing:

```
> library(mice)

mice 2.8 2011-03-24

> md.pattern(weather[,7:10])
    WindGustSpeed Sunshine WindGustDir WindDir9am
329             1        1           1          1  0
  3             1        0           1          1  1
  1             1        1           0          1  1
 31             1        1           1          0  1
  2             0        1           0          1  2
                2        3           3         31 39
```

The table presents, for each variable, a pattern of missing values. Within the table, a 1 indicates a value is present, whereas a 0 indicates a value is missing.

The left column records the number of observations that match the corresponding pattern of missing values. There are 329 observations with no missing values over these four variables (each having a value of 1 within that row). The final column is the number of missing values within the pattern. In the case of the first row here, with no missing values, this is 0.

The rows and columns are sorted in ascending order according to the amount of missing data. Thus, generally, the first row records the number of observations that have no missing values. In our example, the second row corresponds to a pattern of missing values for the variable Sunshine. There are NA hundred NA three observations that have just Sunshine missing (and there are three observations overall that have Sunshine missing based on the final row). This particular row's pattern has just a single variable missing, as indicated by the 1 in the final column.

The final row records the number of missing values over the whole dataset for each of the variables. For example, WindGustSpeed has two missing values. The total number of missing values over all observations and variables is noted at the bottom right (39 in this example).

In Section 7.4, we will discuss how we might deal with missing values through an approach called imputation.

5.2   Visualising Distributions

In the previous section, we purposely avoided any graphical presentation of our data. In fact, I rather expect you might have been frustrated that there was no picture there to help you visualise what we were describing. The absence of a picture was primarily to make the point that it can get a little tricky explaining ideas without the aid of pictures. In particular, our explanation of skewness and kurtosis was quite laboured, and reverted to painting a mental picture rather than presenting an actual picture. After reviewing this current section, go back to reconsider the discussion of skewness and kurtosis. Pictures really do play a significant role in understanding, and graphical presentations, for many, are more effective for communicating than tables of numbers.

Graphical tools allow us to visually investigate the data's characteristics to help us understand it. Such an exploration of the data can clearly identify errors in the data or oddities about its collection. This will also guide our choice of options to transform variables in different ways and to select those variables of interest.

Visualising data has been an area of study within statistics for many years. A vast array of techniques have been developed for presenting data visually, and the topic is covered in great detail in many books, including Cleveland (1993) and Tufte (1985).

It is a good idea, then, early on in a data mining project, to review the distributions of the values of each of the variables in our dataset graphically. R provides a wealth of options for graphically presenting data. Indeed, R is one of the most capable data visualisation languages and allows us to program the visualisations. There are also many standard types of visualisations, and some of these are available through Rattle's Distributions option on the Explore tab (Figure 5.2).

Using Rattle's Distributions option, we can select specific variables of interest and display various distribution plots. Selecting many variables will of course lead to many plots being displayed, and so it may be useful to display multiple plots per page (i.e., per window). Rattle will do this for us automatically, controlled by our setting of the appropriate value for the number of plots per page within the interface. By default, four plots are displayed per page or window.

Figure 5.3 illustrates a sample of the variety of plots available. Clockwise from the top left plot, we have illustrated a **box plot**, a **histogram**, a **mosaic plot**, and a **cumulative function plot**. Because we have

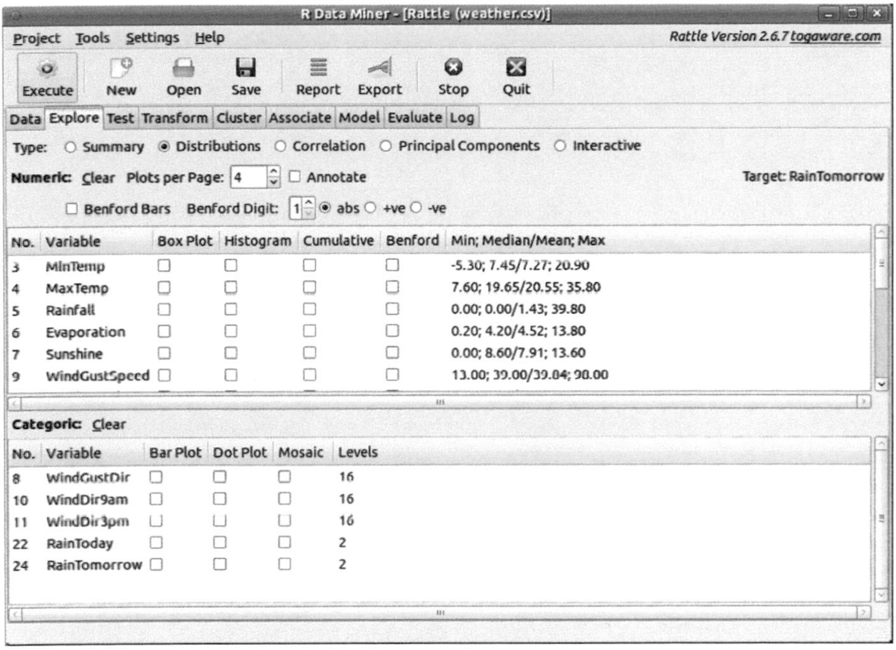

Figure 5.2: The Explore tab's Distributions option provides convenient access to a variety of standard plots for the two primary variable types—numeric and categoric.

identified a target variable (RainTomorrow), the plots include the distributions for each subset of observations associated with each value (No and Yes) of the target variable. That is, the plots include a visualisation of the stratification of the data based on the different values of the target variable.

In brief, the box plot identifies the median and mean of the variable (MinTemp) and the spread from the first quartile to the third, and indicates the outliers. The histogram splits the range of values of the variable (Sunshine) into segments (hours in this case) and shows the number of observations in each segment. The mosaic plot shows the proportions of data split according to the target (RainTomorrow) and the chosen variable (WindGustDir, modified to have fewer levels in this case). The cumulative plot shows the percentage of observations below any particular value of the variable (WindGustSpeed).

Each of the plots available through Rattle is explained in more detail in the following sections.

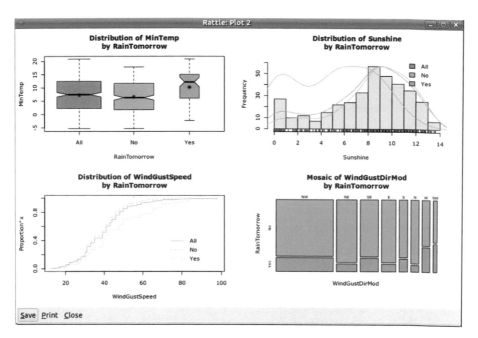

Figure 5.3: A sample of plots illustrates the different distributions and how they can be visualised.

5.2.1 Box Plot

A box plot (Tukey, 1977) (also known as a box-and-whisker plot) provides a graphical overview of how the observations of a variable are distributed. Rattle's box plot adds some additional information to the basic box plot provided by R.

A box plot is useful for quickly ascertaining the distribution of numeric data, covering some of the same statistics presented textually in Section 5.1.1. In particular, any skewness will be clearly visible.

When a target variable has been identified the box plot will also show the distribution of the observations of the chosen variable by the levels of the target variable. We see such a plot for the variable Humidity3pm in Figure 5.4, noting that RainTomorrow is the target variable. The width of each of the box plots also indicates the distribution of the values of the target variable. We see that there are quite a few more observations with No for RainTomorrow than with Yes.

The box plot (which is shown with Rattle's Annotate option active in Figure 5.4) presents a variety of statistics. The thicker horizontal

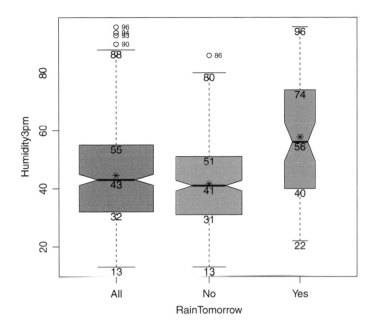

Figure 5.4: The Rattle box plot extends the default R box plots to provide a little more information by default and includes annotations if requested. The plot here is for the full dataset.

line within the box represents the median (also known as the second quartile or the 50th percentile). The leftmost box plot in Figure 5.4 (showing the distribution over all of the observations for Humidity3pm) has the median labelled as 43. The top and bottom extents of the box (55 and 32, respectively) identify the upper quartile (also known as the third quartile or the 75th percentile) and the lower quartile (the first quartile or the 25th percentile). The extent of the box is known as the interquartile range $(55 - 32 - 23)$.

Dashed lines extend to the maximum and minimum data points, which are no more than 1.5 times the interquartile range from the median. We might expect most of the rest of the observations to be within this region. Outliers (points further than 1.5 times the interquartile range from the median) are then individually plotted (we can see a small number of outliers for the left two box plots, each being annotated with the actual value of the observation).

The notches in the box, around the median, indicate an approximate 95% confidence level for the differences between the medians (assuming independent observations, which may not be the case). Thus they are useful in comparing the distributions. In this instance, we can observe that the median of the values associated with the observations for which it rained tomorrow (i.e., the variable `RainTomorrow` has the value `Yes`) is significantly different (at the 95% level of confidence) from the median for those observations for which it did not rain tomorrow. It would appear that a higher humidity recorded at 3 pm is an indication that it might rain tomorrow.

The mean is also displayed as the asterisk in each of the boxes. A large gap between the median and the mean is another indication of a skewed distribution.

Rattle's **Log** tab records the sequence of commands used to draw the box plot and to annotate it. Basically, `boxplot()` (the basic plot), `points()` (to plot the means), and `text()` (to label the various points) are employed.

We can, as always, copy-and-paste these commands into the R Console to replicate the plot and to then manually modify the plot commands to suit any specific need. The automatically generated code is shown below, modified slightly for clarity.

The first step is to generate the data we wish to plot. The following example creates a single dataset with two columns, one being the observations of `Humidity3pm` and the other, identified by a variable called `grp`, the group to which the observation belongs. There are three groups, two corresponding to the two values of the target variable and the other covering all observations.

The use of `with()` allows the variables within the original dataset to be referenced without having to name the dataset each time. We combine three `data.frame()` objects row-wise, using `rbind()`, to generate the final dataset:

```
> ds <- with(crs$dataset[crs$train,],
      rbind(data.frame(dat=Humidity3pm,
                      grp="All"),
            data.frame(dat=Humidity3pm[RainTomorrow=="No"],
                      grp="No"),
            data.frame(dat=Humidity3pm[RainTomorrow=="Yes"],
                      grp="Yes")))
```

Now we display the `boxplot()`, grouping our data by the variable `grp`:

```
> bp <- boxplot(formula=dat ~ grp, data=ds,
               col=rainbow_hcl(3),
               xlab="RainTomorrow", ylab="Humidity3pm",
               notch=TRUE)
```

Notice that we assign to the variable bp the value returned by `boxplot()`. The function returns the data for the calculation needed to draw the box plot. By saving the result, we can make further use of it, as we do below, to annotate the plot.

We will also annotate the plot with the means. To do so, `summaryBy()` from **doBy** comes in handy. The use of `points()` together with pch= results in the asterisks we see in Figure 5.4.

```
> library(doBy)
> points(x=1:3, y=summaryBy(formula=dat ~ grp, data=ds,
               FUN=mean, na.rm=TRUE)$dat.mean, pch=8)
```

Next, we add further `text()` annotations to identify the median and interquartile range:

```
> for (i in seq(ncol(bp$stats)))
  {
    text(x=i, y=bp$stats[,i] -
         0.02*(max(ds$dat, na.rm=TRUE) -
               min(ds$dat, na.rm=TRUE)),
         labels=bp$stats[,i])
  }
```

The outliers are then annotated using `text()`, but decreasing the font size using cex=:

```
> text(x=bp$group+0.1, y=bp$out, labels=bp$out, cex=0.6)
```

To round out our plot, we add a `title()` to include a main= and a sub= title. We `format()` the current date and time (`Sys.time()`) and include the current user (obtained from `Sys.info()`) in the titles:

```
> title(main="Distribution of Humidity3pm (sample)",
        sub=paste("Rattle",
          format(Sys.time(), "%Y-%b-%d %H:%M:%S"),
          Sys.info()["user"]))
```

A variation of the box plot is the box-percentile plot. This plot provides more information about the distribution of the values. We can see
such a plot in Figure 5.5, which is generated using bpplot() of **Hmisc**.
The following code will generate the plot (at the time of writing this
book, box-percentile plots are not yet available in Rattle):

```
> library(Hmisc)
> h3 <- weather$Humidity3pm
> hn <- h3[weather$RainTomorrow=="No"]
> hy <- h3[weather$RainTomorrow=="Yes"]
> ds <- list(h3, hn, hy)
> bpplot(ds, name=c("All", "No", "Yes"),
         ylab="Humidity3pm", xlab="RainTomorrow")
```

The width within each box (they aren't quite boxes as such, but we get
the idea) is determined to be proportional to the number of observations
that are below (or above) that point. The median and the 25th and 75th
percentiles are also shown.

5.2.2 Histogram

A histogram provides a quick and useful graphical view of the spread of
the data. We can very quickly get a feel for the distribution of our data,
including an idea of its skewness and kurtosis. Histograms are probably
one of the more common ways of visually presenting data.

A histogram plot in Rattle includes three components, as we see in
Figure 5.6. The first of these is obviously the vertical bars. The continuous data in the example here (the wind speed at 9 am) has been
partitioned into ranges, and the frequency of each range is displayed as
the bar. R automatically chooses both the partitioning and how the x-
axis is labelled, showing x-axis points at 0, 10, 20, and so on. We might
observe that the most frequent range of values is in the 4–6 partition.

The plot also includes a line plot showing the so-called density estimate. The density plot is a more accurate display of the actual (at least
estimated true) distribution of the data (the values of WindSpeed9am).

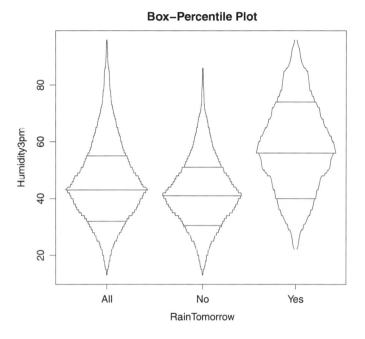

Figure 5.5: A box-percentile plot provides some more information about the distribution.

It allows us to see that rather than values in the range 4–6 occurring frequently, in fact it is "6" itself that occurs most frequently.

The third element of the plot is the so-called *rug* along the bottom of the plot. The rug is a single-dimensional plot of the data along the number line. It is useful in seeing exactly where data points actually lie. For large collections of data with a relatively even spread of values, the rug ends up being quite black. From Figure 5.6, we can make some observations about the data. First, it is clear that the measure of wind speed is actually an integer. Presumably, in the source data, it is rounded to the nearest integer. We can also observe that some values are not represented at all in the dataset. In particular, we can see that 0, 2, 4, 6, 7, and 9 are represented in the data but 1, 3, 5, and 8 are not.

The distribution of the values for WindSpeed9am is also clearly skewed, having a longer tail to the right than to the left. Recall from Section 5.1.4 that WindSpeed9am had a skewness of 1.36. Similarly, the kurtosis measure was 1.48, indicating a bit of a narrower peak.

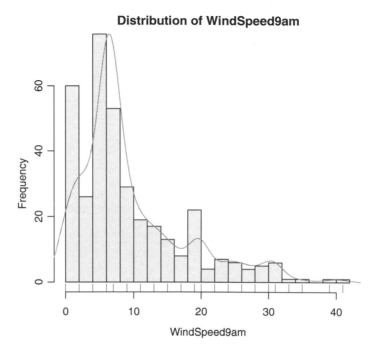

Figure 5.6: The Rattle histogram extends the default R histogram plots with a density plot and a rug plot.

We can compare WindSpeed9am with Sunshine, as in Figure 5.7. The corresponding skewness and kurtosis for Sunshine are −0.72 and −0.27, respectively. That is, Sunshine has a smaller and negative skew, and a smaller kurtosis and hence a more spread-out peak.

5.2.3   Cumulative Distribution Plot

Another popular plot for communicating the distribution of the values of a variable is the cumulative distribution plot. A cumulative distribution plot displays the proportion of the data that has a value that is less than or equal to the value shown on the x-axis.

Figure 5.8 shows a cumulative distribution plot for two variables, WindSpeed9am and Sunshine. Each chart includes three cumulative plots: one line is drawn for all the data and one line for each of the values of the target variable.

We can see again that these two variables have quite different distri-

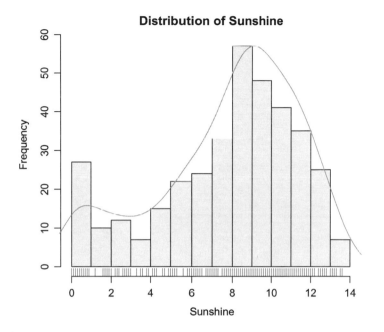

Figure 5.7: The Rattle histogram for Sunshine for comparison with Wind-Speed9am.

butions. The plot for WindSpeed9am indicates that the wind speed at 9 am is usually at the lower end of the scale (e.g., less than 10), but there are a few days with quite extreme wind speeds at 9 am (i.e., outliers). For Sunshine there is a lot more data around the middle, which is typical of a more normal type of distribution. There is quite a spread of values between 6 and 10.

The Sunshine plot is also interesting. We can see quite an obvious difference between the two lines that represent All of the observations and just those with a No (i.e., observations for which there is no rain tomorrow) and the line that represents the Yes observations. It would appear that lower values of Sunshine today are associated with observations for which it rains tomorrow.

The Ecdf() command of **Hmisc** provides a simple interface for producing cumulative distribution plots. The code to generate the Sunshine plot is presented below.

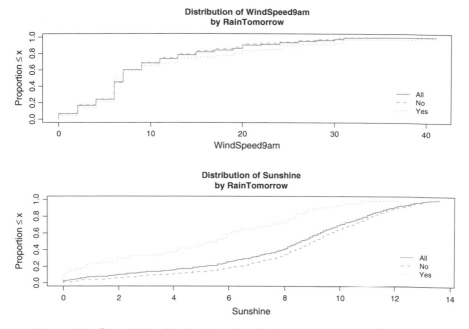

Figure 5.8: Cumulative distribution plots for WindSpeed9am and Sunshine.

```
> library(rattle)
> library(Hmisc)
> su <- weather$Sunshine
> sn <- su[weather$RainTomorrow=="No"]
> sy <- su[weather$RainTomorrow=="Yes"]
> Ecdf(su, col="#E495A5", xlab="Sunshine", subtitles=FALSE)
> Ecdf(sn, col="#86B875", lty=2, add=TRUE, subtitles=FALSE)
> Ecdf(sy, col="#7DB0DD", lty=3, add=TRUE, subtitles=FALSE)
```

We can add a legend and a title to the plot:

```
> legend("bottomright", c("All","No","Yes"), bty="n",
        col=c("#E495A5", "#86B875", "#7DB0DD"),
        lty=1:3, inset=c(0.05,0.05))
> title(main=paste("Distribution of Sunshine (sample)",
        "by RainTomorrow", sep="\n"),
       sub=paste("Rattle", format(Sys.time(),
        "%Y-%b-%d %H:%M:%S")))
```

5.2.4   Benford's Law

The use of Benford's law has proven to be effective in identifying oddities in data. It has been used for case selection in fraud detection, particularly in accounting data (Durtschi et al., 2004), where the value of a variable for a group of related observations might be identified as not conforming to Benford's law even though other groups do.

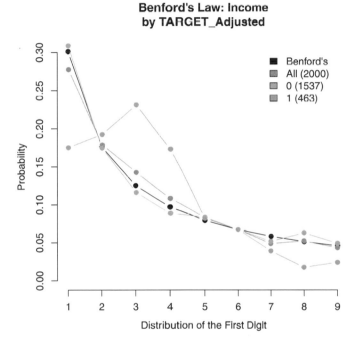

Figure 5.9: A Benford's law plot of the variable Income from the *audit* dataset, particularly showing nonconformance for the population of known noncompliant clients.

Benford's law relates to the frequency of occurrence of the first digit in a collection of numbers. These numbers might be the dollar income earned by individuals across a population of taxpayers or the height of buildings in a city. The law generally applies when several orders of magnitude (e.g., 10, 100, and 1000) are recorded in the observations.

The law states that the digit "1" appears as the first digit of the numbers some 30% of the time. That is, for income, numbers like $13,245 and $162,385 (having an initial digit of "1") will appear about 30% of the time in our population. On the other hand, the digit "9" (for example, as in $94,251) appears as the first digit less than 5% of the time. Other digits have frequencies between these two, as we can see from the black line in Figure 5.9.

This rather startling observation is certainly found, empirically, to hold in many collections of numbers, such as bank account balances, tax refunds, stock prices, death rates, lengths of rivers, and potential fraud in elections. It is observed to hold for processes that are described by what are called power laws, which are common in nature. By plotting a collection of numbers against the expectation as based on Benford's law, we are able to quickly see any odd behaviour in the data.

Benford's law is not valid for all collections of numbers. For example, peoples' ages would not be expected to follow Benford's law, nor would telephone numbers. So we do need to use caution in relying just on Benford's law to identify cases of interest.

We can illustrate Benford's law using the *audit* dataset from **rattle**. Rattle provides a convenient mechanism for generating a plot to visualise Benford's law, and we illustrate this with the variable Income in Figure 5.9.

The darker line corresponds to Benford's law, and we note that the lines corresponding to All and 0 follow the expected first-digit distribution proposed by Benford's law. However, the line corresponding to 1 (i.e., clients who had to have their claims adjusted) clearly deviates from the proposed distribution. This might indicate that these numbers have been made up by someone or that there is some other process happening that affects this population of numbers.

5.2.5   Bar Plot

The plots we have discussed so far work for numeric data. We now consider plots that work for categoric data. These include the bar plot, dot plot, and mosaic plot.

A bar plot, much like a histogram, uses vertical bars to show counts of the number of observations of each of the possible values of the categoric variable. There are many ways to graph a bar plot. In Rattle, the default is to list the possible values along the x-axis, leaving the y-axis for the frequency or count. When a categoric target variable is active within Rattle, additional bars will be drawn for each value, corresponding to the different values of the target. Figure 5.10 shows a typical bar plot. The sample bar plot also shows that by default Rattle will sort the bars from the largest to the smallest.

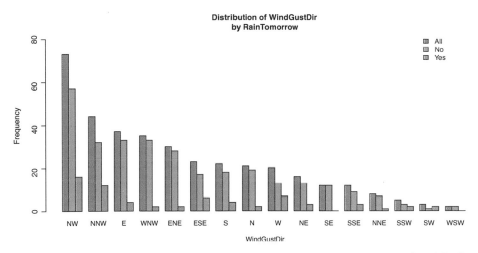

Figure 5.10: A bar plot for the categoric variable WindGustDir (modified) from the *weather* dataset.

5.2.6 Dot Plot

A dot plot illustrates much the same information as a bar plot but uses a different approach. The bars are replaced by dots that show the height, thus not filling in as much of the graphic. A dotted line replaces the extent of the bar, and by default in Rattle the plots are horizontal rather than vertical. Once again, the categoric values are ordered to produce the plot in Figure 5.11.

For the dot plot, we illustrate the distribution of observations over all of the values of the categoric variable WindGustDir. With the horizontal plot it is more feasible to list all of the values than for the vertical bar plots. Of course, the bar plots could be drawn horizontally for the same reason.

Both the bar plot and dot plot are useful in understanding how the observations are distributed across the different categoric values. One thing to look for when a target variable is identified is any specific variation in the distributions between the target values. In Figure 5.11, for example, we can see that the distributions of the "Yes" and "No" observa-

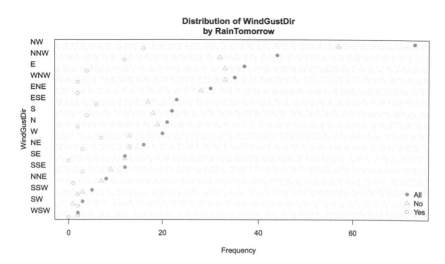

Figure 5.11: A dot plot for the categoric variable WindGustDir (original) from the *weather* dataset.

tions are quite different from the overall distributions. Such observations may merely be interesting or might lead to useful questions about the data. In our data here, we do need to recall that the majority of days have no rain. Thus the No distribution of values for WindGustDir follows the distribution of all observations quite closely. We can see a few deviations, suggesting that these wind directions have an influence on the target variable.

5.2.7   Mosaic Plot

A mosaic plot is an effective way to visualise the distribution of the values of one variable over the different values of another variable, looking for any structure or relationship between those variables. In Rattle, this second variable is usually the target variable (e.g., RainTomorrow in our *weather* dataset), as in Figure 5.12.

The mosaic plot again provides insights into how the data is distributed over the values of a second variable. The area of each bar is proportional to the number of observations having a particular value for the variable WindGustDir. Once again, the values of WindGustDir are ordered according to their frequency of occurrence. The value NW is observed most frequently. The split between the values of the target

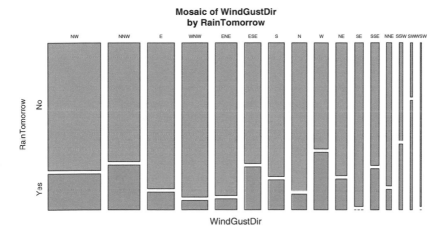

Figure 5.12: A mosaic plot for the categoric variable WindGustDir (original) from the *weather* dataset.

variable RainTomorrow is similarly proportional to their frequency.

Once again, we see that a wind gust of west has a high proportion of days for which RainTomorrow is true. Something that we missed in reviewing the bar and the dot plots is that a SW wind gust has the highest proportions of days where it rains tomorrow, followed by SSW.

It is arguable, though, that it is harder to see the overall distribution of the wind gust directions in a mosaic plot compared with the bar and dot plots. Mosaic plots are thus generally used in combination with other plots, and they are particularly good for comparing two or more variables at the same time.

5.2.8   Pairs and Scatter Plots

The bar and dot plots are basically single-variable (i.e., univariate) plots. In our plots, we have been including a second variable, the target. Moving on from considering the distribution of a single variable at a time, we can compare variables pairwise. Such a plot is called a **scatter plot**.

Generally we have multiple variables that we might wish to compare pairwise using multiple scatter plots. Such a plot then becomes a **scatter plot matrix**. The pairs() command in R can be used to generate a matrix of scatter plots. In fact, the function can be fine-tuned to not only display pairwise scatter plots but also to include histograms and a pairwise measure of the correlation between variables (correlations are discussed in Section 5.3).

For this added functionality we need two support functions that are a little more complex and that we won't explain in detail:

```
> panel.hist <- function(x, ...)
  {
    usr <- par("usr"); on.exit(par(usr))
    par(usr=c(usr[1:2], 0, 1.5) )
    h <- hist(x, plot=FALSE)
    breaks <- h$breaks; nB <- length(breaks)
    y <- h$counts; y <- y/max(y)
    rect(breaks[-nB], 0, breaks[-1], y, col="grey90", ...)
  }
> panel.cor <- function(x, y, digits=2, prefix="",
                        cex.cor, ...)
  {
    usr <- par("usr"); on.exit(par(usr))
    par(usr = c(0, 1, 0, 1))
    r <- (cor(x, y, use="complete"))
    txt <- format(c(r, 0.123456789), digits=digits)[1]
    txt <- paste(prefix, txt, sep="")
    if(missing(cex.cor)) cex.cor <- 0.8/strwidth(txt)
    text(0.5, 0.5, txt)
  }
```

We can then generate the plot with:

```
> vars <- c(5, 7, 8, 9, 15, 24)
> pairs(weather[vars],
        diag.panel=panel.hist,
        upper.panel=panel.smooth,
        lower.panel=panel.cor)
```

There are two additional commands defined here, panel.hist() and panel.cor(), provided as the arguments diag.panel and lower.panel to pairs(). These two commands are not provided by R directly. Their definitions can be obtained from the help page for pairs() and pasted into the R Console.

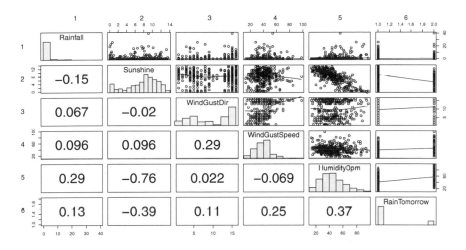

Figure 5.13: A pairs plot with a scatter plot matrix displayed in the upper panel, histograms in the diagonal, and a measure of correlation in the lower panel.

> **Tip:** *Rattle can generate a pairs plot including the scatter plots, histograms, and correlations in the one plot (Figure 5.13). To do so, we go to the* **Distributions** *option of the* **Explore** *tab and ensure that no plot types are selected for any variable (the default). Then click the* **Execute** *button.*

Notice that we have only included six variables in the pairs plot. Any more than this and the plot becomes somewhat crowded. In generating a pairs plot, Rattle will randomly subset the total number of variables available down to just six variables. In fact, each time the Execute button is clicked, a different randomly selected collection of variables will be displayed. This is a useful exercise to explore for interesting pairwise relationships among our variables. If we are keen to do so, we can generate plots with more than six variables quite simply by copying the command from Rattle's Log tab (which will be similar to pairs(), shown above) and pasting it into the R Console.

Let's explore the pairs plot in a little more detail. The diagonal contains a histogram for the numeric variables and a bar plot (also a histogram) for the categoric variables. The top right plots (i.e., those plots above the diagonal) are pairwise scatter plots, which plot the observations of just two variables at a time. The corresponding variables are identified from the diagonal.

The top left scatter plot, which appears in the first row and the second column, has `Rainfall` on the y-axis and `Sunshine` on the x-axis. We can see quite a predominance of days (observations) with what looks like no rainfall at all, with fewer observations having some rainfall. There does not appear to be any particular relationship between the amount of rain and the hours of sunshine, although there are some days with higher rainfall when there is less sunshine. Note, though, that an outlier for rainfall (at about 40 mm of rain) appears on a day with about 9 hours of sunshine. We might decide to explore this apparent anomaly to assure ourselves that there was no measurement or data error that led to this observation.

An interesting scatter plot to examine is that in row 2 and column 5. This plot has `Sunshine` on the y-axis and `Humidity3pm` on the x-axis. The solid red lines that are drawn on the plot are a result of `panel.smooth()` being provided as the value for the `upper.panel` argument in the call to `pairs()`. The line provides a hint of any trend in the relationship between the two variables. For this particular scatter plot, we can see some structure in that higher levels of humidity at 3 pm are observed with lower hours of sunshine. One or two of the other scatter plots show other, but less pronounced and hence probably less significant, relationships.

The lower part of our scatter plot matrix contains numbers between −1 and 1. These are measures of the correlation between two variables. Pearson's correlation coefficient is used. We can see that `Rainfall` and `Humidity3pm` (see the number in row 5, column 1) have a small positive correlation of 0.29. That is not a great deal of correlation. If we square the correlation value to obtain 0.0841, we can interpret this as indicating that some 8% of the variation is related. There is perhaps some basis to expect that when we observe higher rainfall we might also observe higher humidity at 3 pm.

There is even stronger correlation between the variables `Sunshine` and `Humidity3pm` (row 5, column 2) measured at −0.76. The negative sign indicates a negative correlation of strength 0.76. Squaring this number leads us to observe that some 58% of the variation is related. Thus, observations of more sunshine do tend to occur with observations of less humidity at 3 pm, as we have already noted. We will come back to correlations shortly.

5.2.9 Plots with Groups

Extending the idea of comparing variables, we can usefully plot, for example, a box plot of one variable but with the observations split into groups that are defined by another variable. We have already seen this with the target variable being the one by which we group our observations, as in Figure 5.4. Simply through selecting another variable as the target, we can explore many different relationships quite effectively.

Consider, for example, the distribution of the observations of the variable Sunshine. We might choose Cloud9am as the target variable (in Rattle's Data tab) and then request a box plot from the Explore tab. The result will be as in Figure 5.14. Note that Cloud9am is actually a numeric variable, but we are effectively using it here as a categoric variable. This is okay since it has only nine numeric values and those are used here to group together different observations (days) having a common value for the variable.

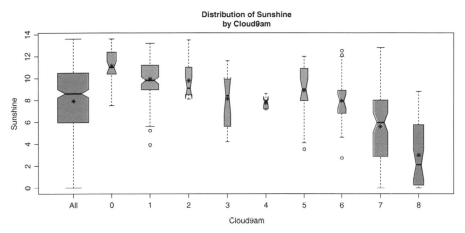

Figure 5.14: Data displayed through the Distributions tab is grouped using the target variable values to define the groups. Selecting alternative targets will group the data differently.

The leftmost box plot shows the distribution of the observations of Sunshine over the whole dataset. The remaining nine box plots then collect together the observations of Sunshine for each of the nine possible values of Cloud9am. Recall that Cloud9am is measured in something called oktas. An okta of 0 indicates no cloud coverage, 1 indicates one-eighth of the sky is covered, and so on up to 8, indicating that the sky is completely covered in clouds.

The relationship that we see in Figure 5.14 then makes some sense. There is a clear downward trend in the box plots for the amount of sunshine as we progressively have more cloud coverage. Some groups are quite distinct: compare groups 6, 7, and 8. They have different medians, with their notches clearly not overlapping.

The plot also illustrates some minor idiosyncrasies of the box plot. The box plots for groups 2 and 4 appear a little different. Each has an odd arrangement at the end of the quartile on one side of the box. This occurs when the notch is calculated to be larger than the portion of the box on one side of the median.

5.3   Correlation Analysis

We have seen from many of the plots in the sections above, particularly those plots with more than a single variable, that we often end up identifying some kind of relationship or correlation between the observations of two variables. The relationship we saw between `Sunshine` and `Humidity3pm` in Figure 5.13 is one such example.

A correlation coefficient is a measure of the degree of relationship between two variables—it is usually a number between −1 and 1. The magnitude represents the strength of the correlation and the sign represents the direction of the correlation. A high degree of correlation (closer to 1 or −1) indicates that the two variables are very highly correlated, either positively or negatively. A high positive correlation indicates that observations with a high value for one variable will also tend to have a high value for the second variable. A high negative correlation indicates that observations with a high value for one variable will also tend to have a lower value of the second variable. Correlations of 1 (or −1) indicate that the two variables are essentially identical, except perhaps for scale (i.e., one variable is just a multiple of the other).

5.3.1   Correlation Plot

From our previous exploration of the *weather* dataset, we noted a moderate (negative) correlation between `Sunshine` and `Humidity3pm`. Generally, days with a higher level of sunshine have a lower level of humidity at 3 pm and vice versa.

Variables that are very strongly correlated are probably not independent. That is, they have some close relationship. The relationship could be causal in that an increase in one has some physical impact on the other. But such evidence for this needs to be ascertained separately. Nonetheless, having correlated variables as input to some algorithms may misguide the data mining. Thus it is important to recognise this.

R can be used to quite easily generate a matrix of correlations between variables. The cor() command will calculate and list the Pearson correlation between variables:

```
> vars <- c(5, 6, 7, 9, 15)
> cor(weather[vars], use="pairwise", method="pearson")

                Rainfall Evaporation Sunshine
Rainfall        1.000000   -0.007293 -0.15099
Evaporation    -0.007293    1.000000  0.31803
Sunshine       -0.150990    0.318025  1.00000
WindGustSpeed   0.096190    0.288477  0.09584
Humidity3pm     0.289013   -0.391780 -0.75943
                WindGustSpeed Humidity3pm
Rainfall             0.09619      0.28901
Evaporation          0.28848     -0.39178
Sunshine             0.09584     -0.75943
WindGustSpeed        1.00000     -0.06944
Humidity3pm         -0.06944      1.00000
```

We can compare these numbers with those in Figure 5.13. They should agree. Note that each variable is, of course, perfectly correlated with itself, and that the matrix here is symmetrical about the diagonal (i.e., the measure of the correlation between Rainfall and Sunshine is the same as that between Sunshine and Rainfall).

We have to work a little hard to find patterns in the matrix of correlation values expressed in this way. Rattle provides access to a graphical plot of the correlations between variables in our dataset. The Correlation option of the Explore tab provides a number of choices for correlation plots (Figure 5.15). Simply clicking the Execute button will cause the default correlation plot to be displayed (Figure 5.16).

The first thing we might notice about this correlation plot is that only the numeric variables appear. Rattle only computes correlations between numeric variables. The second thing to note about the plot is

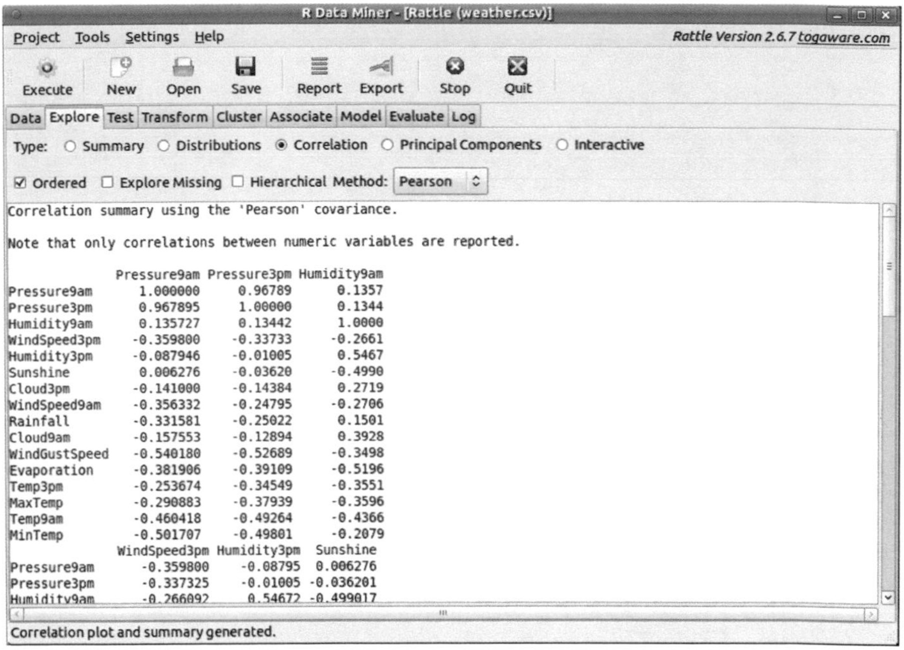

Figure 5.15: The Explore tab's Correlation option provides access to plots that visualise correlations between pairs of variables.

that it is symmetric about the diagonal, as is the numeric correlation matrix we saw above—the correlation between two variables is the same, irrespective of the order in which we view them. The third thing to note is that the order of the variables does not correspond to the order in the dataset but to the order of the strength of any correlations, from the least to the greatest. This is done to achieve a more pleasing graphic but can also lead to further insight with groupings of similar correlations. This is controlled through the **Ordered** check button.

We can understand the degree of any correlation between two variables by both the shape and the colour of the graphic elements. Any variable is, of course, perfectly correlated with itself, and this is reflected as the straight lines on the diagonal of the plot.

A perfect circle, on the other hand, indicates that there is no (or very little) correlation between the variables. This appears to be the case, for example, for the correlation between `Sunshine` and `Pressure9am`. In fact, there is a correlation, just an extremely weak one (0.006), as we see in Figure 5.15.

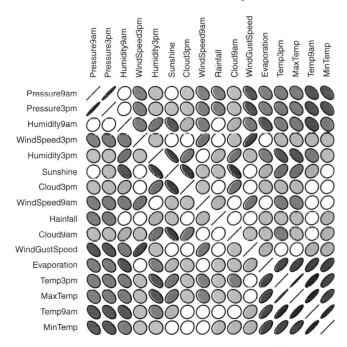

Figure 5.16: The correlation plot graphically displays different degrees of correlation pairwise between variables.

The circles turn into straight lines, by degrees, as the strength of correlation between the two variables increases. Thus we can see that there is some moderate correlation between Humidity9am and Humidity3pm, represented as the squashed circle (i.e., an ellipse shape). The more squashed (i.e., the more like a straight line), the higher the degree of correlation, as in the correlation between MinTemp and Temp9am. Notice that, intuitively, all of the observations of correlations make some sense.

The direction of the ellipse indicates whether the correlation is positive or negative. The correlations we noted above were in the positive direction. We can see, for example, our previously observed negative correlation between Sunshine and Humidity3pm.

The colours used to shade the ellipses give another, if redundant, clue to the strength of the correlation. The intensity of the colour is maximal (black) for a perfect correlation and minimal (white) if there is no correlation. Shades of red are used for negative correlations and blue for positive correlations.

5.3.2   Missing Value Correlations

An interesting and useful twist on the concept of correlation analysis is the concept of correlation amongst missing values in our data. In many datasets, it is often constructive to understand the nature of missing data. We often find commonality amongst observations with a missing value for one variable having missing values for other variables. A correlation plot can effectively highlight such structure in our datasets.

The correlation between missing values can be explored by clicking the Explore Missing check box. To understand missing values fully, we have also turned off the partitioning of the dataset on the Data tab so that all of the data is considered for the plot. The resulting plot is shown in Figure 5.17.

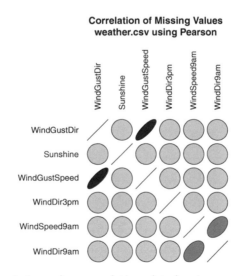

Figure 5.17: The missing values correlation plot showing correlations between missing values of variables.

We notice immediately that only six variables are included in this correlation plot. Rattle has identified that the other variables have no missing values, and so there is no point including them in the plot. We also notice that a categoric variable, WindGustDir, is included in the plot even though it was not included in the usual correlation plot. We can obtain a correlation for categoric variables since we only measure the absence or presence of a value, which is easily interpreted as numeric.

The graphic shows us that WindGustSpeed and WindGustDir are quite highly correlated with respect to missing values. That is, when the variable WindGustSpeed has a missing value, WindGustDir also tends to have a missing value, and vice versa. The actual correlation is 0.81489 (which can be read from the Rattle text view window). There is also a weak correlation between WindDir9am and WindSpeed9am (0.36427).

On the other hand, there is no (in fact, very little at −0.0079) correlation between Sunshine and WindGustSpeed, or any other variable, with regard to missing values.

It is important to note that the correlations showing missing values may be based on very small samples, and this information is included in the text view of the Rattle window. For example, in this case we can see in Figure 5.18 that there are only 21 missing observations for WindDir9am and only two or three for the other variables. This corresponds to approximately 8% and 1% of the observations, respectively, having missing values for these variables. This is too little to draw too many conclusions from.

5.3.3   Hierarchical Correlation

Another useful option provided by Rattle is the hierarchical correlation plot (Figure 5.19). The plot provides an overview of the correlation between variables using a tree-like structure known as a dendrogram.

The plot lists the variables in the right column. The variables are then linked together in the dendrogram according to how well they are correlated. The x-axis is a measure of the height within the dendrogram, ranging from 0 to 3. The heights (i.e., lengths of the lines within the dendrogram) give an indication of the level of correlation between variables, with shorter heights indicating stronger correlations.

Very quickly we can observe that Temp3pm and MaxTemp are quite closely correlated (in fact, they have a correlation of 0.99). Similarly, Cloud3pm and Cloud9am are moderately correlated (0.51). The group of variables Temp9am, MinTemp, Evaporation, Temp3pm, and MaxTemp, unsurprisingly, have some higher level of correlation amongst themselves than they do with other variables.

A number of R functions are used together to generate the plot we see in Figure 5.19. We take the opportunity to review the R code to gain

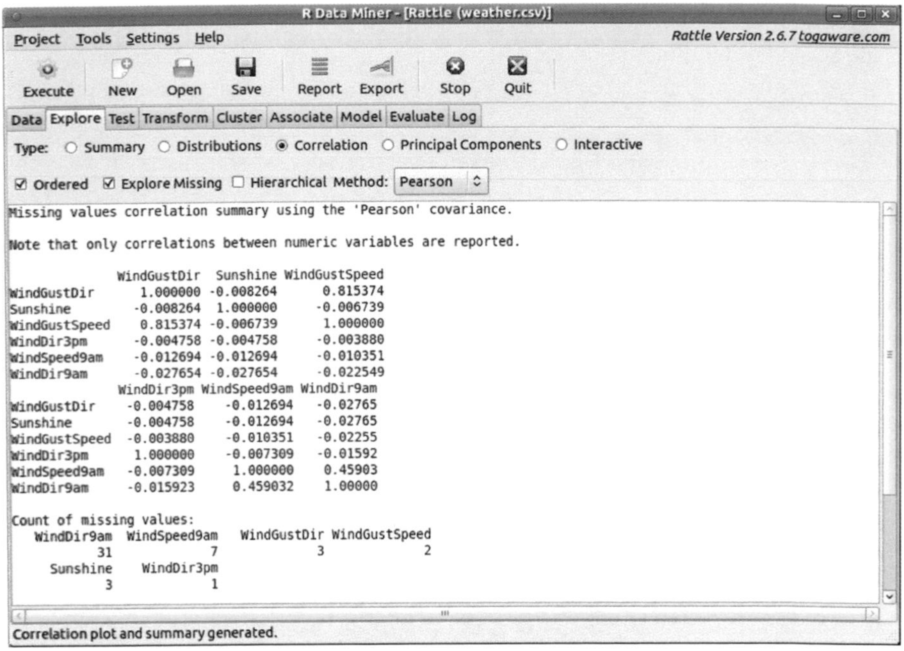

Figure 5.18: The Rattle window displays the underlying data used for the missing observations correlation plot.

a little more understanding of working directly with R. Rattle's Log tab will again provide the steps, which include generating the correlations for the numeric variables using cor():

```
> numerics <- c(3:7, 9, 12:21)
> cc <- cor(weather[numerics],
            use="pairwise",
            method="pearson")
```

We then generate a hierarchical clustering of the correlations. This can be done using hclust() (cluster analysis is detailed in Chapter 9):

```
> hc <- hclust(dist(cc), method="average")
```

A dendrogram, the graph structure that we see in the plot for Figure 5.19, can then be constructed using as.dendrogram():

```
> dn <- as.dendrogram(hc)
```

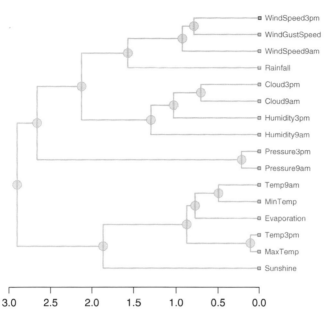

Variable Correlation Clusters
weather.csv using Pearson

Figure 5.19: The Hierarchical option displays the correlation between variables using a dendrogram.

The actual plot is drawn using `plot()`:

```
> plot(dn, horiz = TRUE)
```

5.4   Command Summary

This chapter has referenced the following R packages, commands, functions, and datasets:

audit	dataset	Used to illustrate Benford's law.
basicStats()	command	Detailed statistics of data.
bpplot()	command	Box-percentile plot.

describe()	command	Detailed data summary.
cor()	function	Correlation between variables.
Ecdf()	command	Produce cumulative distribution plot.
fBasics	package	More comprehensive basic statistics.
hclust()	function	A hierarchical clustering algorithm.
Hmisc	package	Additional basic statistics and plots.
kurtosis()	function	A measure of distribution peakiness.
md.pattern()	command	Table of patterns of missing values.
mice	package	Missing data analysis.
pairs()	command	Matrix of pairwise scatter plots.
panel.hist()	command	Draw histograms within a pairs plot.
panel.cor()	command	Correlations within a pairs plot.
panel.smooth()	command	Add smooth line to pairs plot.
sample()	function	Select a random sample of a dataset.
skewness()	function	A measure of distribution skew.
summary()	command	Basic dataset statistics.
weather	dataset	Sample dataset from **rattle**.

Chapter 6

Interactive Graphics

There is more to exploring data than simply generating textual and statistical summaries and graphical plots. As we have begun to see, R has some very significant capabilities for generating graphics that assist in revealing the story our data is telling us and then helps us to effectively communicate that story to others. However, R is specifically suited to generating static graphics—that is, as Wickham (2009) says, "there is no benefit displaying on a computer screen as opposed to on a piece of paper" when using R's graphics capabilities.

R graphics were implemented with the idea of presenting the data visually rather than interacting with it. We write scripts for the display. We then go back to our script to fine-tune or explore different options for the displayed data. This is great for repeatable generation of graphics but not so efficient for the "follow your nose" or "ad hoc reporting" approach to quick and efficient data exploration.

Being able to easily interact with a plot can add significantly to the efficiency of our data exploration and lead to the discovery of interesting and important patterns and relationships. Data miners will need sophisticated skills in dynamically interacting with the visualisations of data to provide themselves with significant insights. Whilst software supports this to some extent, the true insights come from the skill of the data miner. We must take time to explore our data, identify relationships, discover patterns, and understand the picture painted by the data.

Rattle provides access to two very powerful R packages for interactive data analysis, **latticist** (Andrews, 2010) and GGobi, the latter of which is accessed via **rggobi** (Lang et al., 2011). These can be initiated through the Interactive option of the Explore tab (Figure 6.1). We will introduce each of the tools in this chapter. Note that each application has much more functionality than can be covered here, and indeed **GGobi** has its own book (Cook and Swayne, 2007), which provides good details.

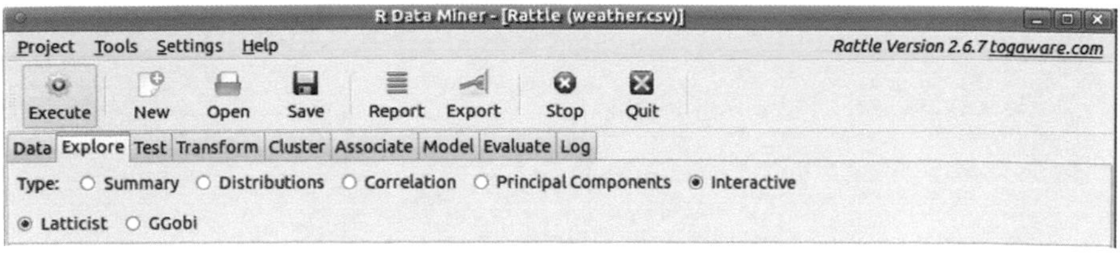

Figure 6.1: The Explore tab's Interactive option can initiate a **latticist** or GGobi session for interactive data analysis.

6.1   Latticist

Latticist (Andrews, 2010) provides a graphical and interactive interface to the advanced plotting capabilities of R's **lattice** (Sarkar, 2008). It is written in R itself and allows the underlying R commands that generate the plots to be directly edited and their effect immediately viewed. This then provides a more interactive experience with the generation of R plots. Select the Latticist radio button of the Interactive option of the Explore tab and then click the toolbar's Execute button to display **latticist**'s window, as shown in Figure 6.2.

From the R Console, we can use latticist() to display the same interactive window for exploring the *weather* dataset:

```
> library(latticist)
> latticist(weather)
```

With the initial Latticist window, we immediately obtain an overall view of some of the story from our data. Note that, by default, from Rattle, the plots show the data grouped by the target variable RainTomorrow. We see that numeric data is illustrated with a density plot, whilst categoric data is displayed using dot plots.

Many of the plots show differences in the distributions for the two groups (based on whether RainTomorrow is No or Yes). We might note, for example, that variables MinTemp and MaxTemp (the first two plots of the top row) have slightly higher values for the observations where it

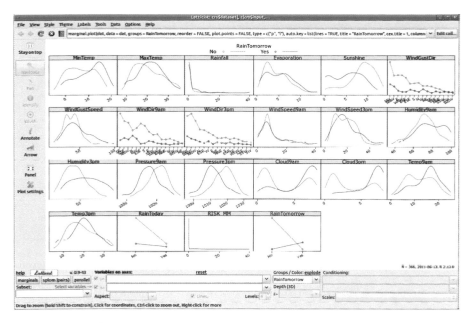

Figure 6.2: The Explore tab's Interactive option can initiate a **latticist** session for interactive data analysis..

rains tomorrow. The third plot suggests that the amount of `Rainfall` today seems to be almost identically distributed for observations where it does not rain tomorrow and those where it does. The fifth plot then indicates that there seems to be less `Sunshine` on days prior to days on which it rains.

There is an extensive set of features available for interacting with the visualisations. The actual command used to generate the current plot is shown at the top of the window. We can modify the command and immediately see the result, either by editing the command in place or clicking the Edit call... button. The latter results in the display of a small text window in which the command can be edited. There are buttons in the main window's toolbar to open the help page for the current plot, to reload the plot, and to navigate to previous plots.

The default plot is a plot of the marginal distribution of the variables. The buttons near the bottom left of the window allow us to select between marginal, splom (pairs), and parallel coordinates plots. A splom is a scatter plot matrix similar to that in Section 5.2.8. A parallel coordinates plot draws a line for each observation from one variable to the next, as in

Figure 6.3. Parallel coordinates plots can be quite useful in identifying groups of observations with similar values across multiple variables.

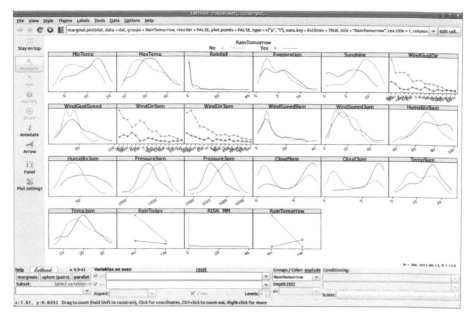

Figure 6.3: The parallel coordinates plot from **latticist**.

The parallel coordinates plot in Figure 6.3 exposes some structure in the *weather* dataset. The top variable in this case is the target variable, `RainTomorrow`. The other variables are `Sunshine`, `Rainfall`, `MaxTemp`, and `MinTemp`. Noting that each line represents a single observation (the weather details for each day), we might observe that for days when there is less sunshine it is more likely to rain tomorrow, and similarly when there is more sunshine it is less likely to rain tomorrow. We can observe a strong band of observations with no rain tomorrow, higher amounts of sunshine today, and little or no rainfall today. From there (to the remaining two variables) we observe less structure in the data.

There is a lot more functionality available in **latticist**. Exploring many of the different options through the interface is fruitful. We can add arrows and text to plots and then export the plots for inclusion in other documents. The data can be subset and grouped in a variety of ways using the variables available. This can lead to many insights, following our nose, so to speak, in navigating our way through the data. All the time we are on the lookout for structure and must remember to capture it to support the story that we find the data telling us.

6.2   GGobi

GGobi is also a powerful open source tool for visualising data, supporting two of the most useful interactive visualisation concepts, known as brushing and tours. GGobi is not R software as such[1] but is integrated with R through **rggobi** (Lang et al., 2011) and ggobi(). Key uses in a data mining context include the exploration of the distribution of observations for multiple variables, visualisations of missing values, exploration for the development of classification models, and cluster analysis. Cook and Swayne (2007) provide extensive coverage of the use of GGobi, particularly relevant in a data mining context.

To use GGobi from the Interactive option of the Explore tab, the GGobi application will need to be installed. GGobi runs under GNU/Linux, Mac OS/X, and Microsoft Windows and is available for download from http://www.ggobi.org/.

GGobi is very powerful indeed, and here we only cover some basic functionality. With GGobi we are able to explore high-dimensional data through highly dynamic and interactive graphics that include tours, scatter plots, bar plots, and parallel coordinates plots. The plots are interactive and linked with brushing and identification. Panning and zooming are supported. Data can be rotated in 3D, and we can tour high-dimensional data through 1D, 2D, and 2x1D projections, with manual and automatic control of projection pursuits.

We are also able to interact with GGobi by issuing commands through the R Console, and thus we can script some standard visualisations from R using GGobi. For example, patterns found in data using R or Rattle can be automatically passed to GGobi for interactive exploration. Whilst interacting with GGobi plots we can also highlight points and have them communicated back to R for further analysis.

Scatter plot

We can start GGobi from Rattle by clicking the Execute button whilst having selected GGobi under the Interactive option of the Explore tab, as in Figure 6.1.

We can also initiate GGobi with rggobi(), providing it with a data frame to load. In this example, we remove the first two variables (Date and Location) and pass on to rggobi() the remaining variables:

---

[1] A project is under way to implement the concepts of GGobi directly in R.

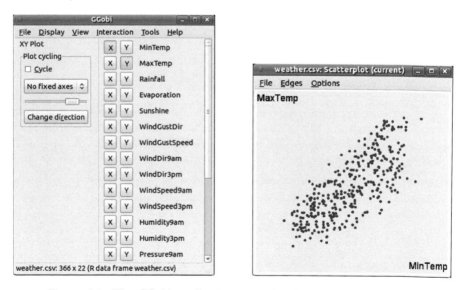

Figure 6.4: The Explore tab's Interactive option can initiate a GGobi session for interactive data analysis. Select GGobi and then click Execute.

```
> library(rggobi)
> gg <- rggobi(weather[-c(1,2)])
```

On starting, GGobi will display the two windows shown in Figure 6.5. The first provides controls for the visualisations and the other displays the default visualisation (a two-variable scatter plot of the first two variables of the data frame supplied, noting that we have removed Date and Location).

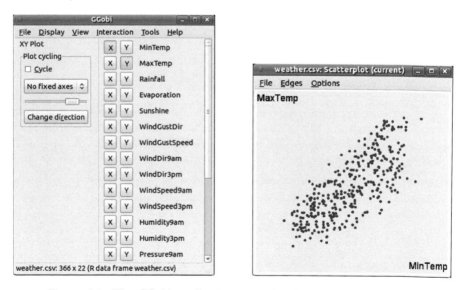

Figure 6.5: The GGobi application control and scatter plot windows.

The control window provides menus to access all of the functionality of GGobi. Below the menu bar, we can currently see the XY Plot (i.e., scatter plot) options. Two variables are selected from the variable list on the right side of the control window. The variables selected for display in the scatter plot are for the x-axis (X) and the y-axis (Y). By default, the first (MinTemp) and second (MaxTemp) are the chosen variables in our dataset. We can choose any of our variables to be the X or the Y by clicking the appropriate button. This will immediately change what is displayed in the plot.

Multiple Plots

Any number of plots can be displayed simultaneously. From the Display menu, we can choose a New Scatterplot Display to have two (or more) plots displayed at one time, each in its own window. Figure 6.6 shows two scatter plots, with the new one chosen to display Evaporation against Sunshine. Changes that we make in the controlling window affect the *current*, plot which can be chosen by clicking the plot. We can also do this from the R Console using display():

```
> display(gg[1], vars=list(X="Evaporation", Y="Sunshine"))
```

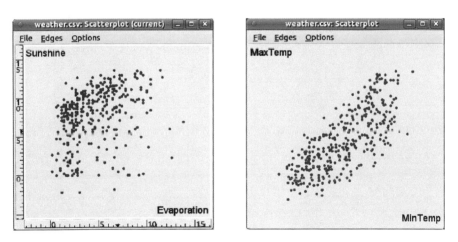

Figure 6.6: Multiple scatter plots from GGobi with and without axes.

Brushing

Brushing allows us to select observations in any plot and see them highlighted in all plots. This lets us visualise across many more dimensions than possible with a single two-dimensional plot.

From a data mining perspective, we are usually most interested in the relationship between the input variables and the target variable (using the variable RainTomorrow in our examples). We can highlight its two different values for the different observations using colour. From the Tools menu, choose Automatic Brushing to display the window shown in Figure 6.7.

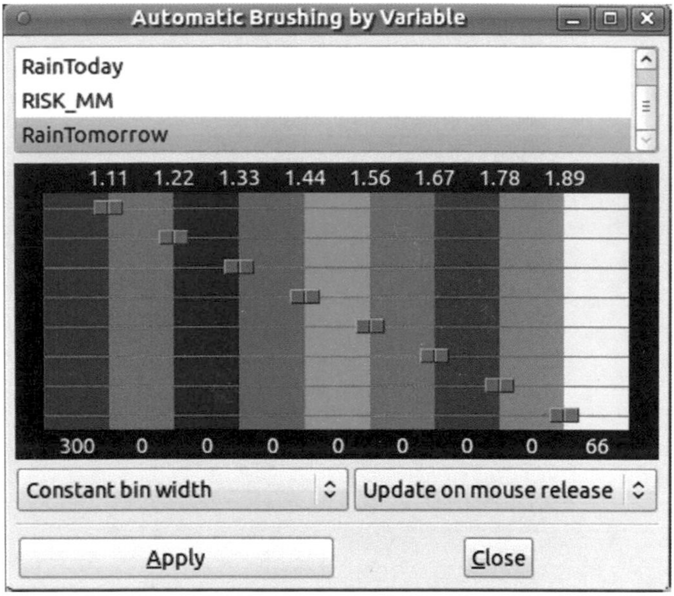

Figure 6.7: GGobi's automatic brushing. The frequencies along the bottom will be different depending on whether or not the data is partitioned within Rattle.

From the list of variables that we see at the top of the resulting window, we can choose RainTomorrow (after scrolling through the list of variables to find RainTomorrow at the bottom of the list). Notice that the number ranges that are displayed in the lower colour map change to reflect the range of values associated with the chosen variable. For RainTomorrow, which has only the values 0 and 1, any observations having RainTomorrow values of 0 will be coloured purple, whilst those with a value of 1 will be coloured yellow.

We click on the Apply button for the automatic brushing to take effect. Any plots that GGobi is currently displaying (and any new plots we cause to be displayed from now on) will colour the observations appropriately, as in Figure 6.8. This colouring of points across multiple plots is referred to as *brushing*.

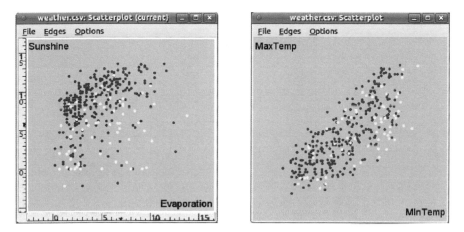

Figure 6.8: Automatic brushing of multiple scatterplots using GGobi.

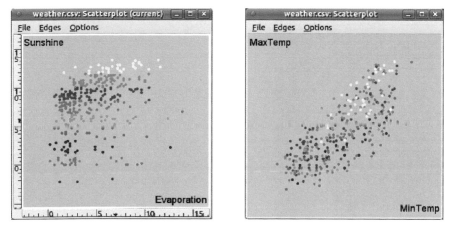

Figure 6.9: Colourful brushing of multiple scatterplots.

Our plots can be made somewhat more colourful by choosing a numeric variable, like Sunshine, as the choice for automatic brushing. We can see the effect in Figure 6.9. GGobi provides an extensive collection of colour schemes to choose from for these gradients, for example. Under the Tools menu, select the Color Schemes option. A nice choice could be Y1OrRd9.

Other Plots

The Display menu provides a number of other options for plots. The Scatterplot Matrix, for example, can be used to display a matrix of scatter plots across many variables at one time. We've seen this already in both Rattle itself and **latticist**. However, GGobi offers brushing and linked views across all of the currently displayed GGobi plots.

By default, the Scatterplot Matrix will display the first four variables in our dataset, as shown in Figure 6.10. We can add and remove variables by selecting the appropriate buttons in the control window, which we notice has changed to include just the Scatterplot Matrix options rather than the previous Scatterplot options. Any manual or automatic brushing in effect will also be reflected in the scatter plots, as we can see in Figure 6.10.

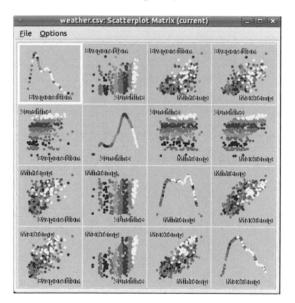

Figure 6.10: GGobi's scatter plot matrix.

A parallel coordinates plot is also easily generated from GGobi's Display menu. An example can be seen in Figure 6.11, showing five variables, beginning with Sunshine. The automatic brushing based on Sunshine is still in effect, and we can see that the coloured lines emanate from the left end of the plot within colour groups. The yellow lines represent observations with a higher value of Sunshine, and we can see that these generally correspond to higher values of the other variables here, except for the final variable (`Rainfall`).

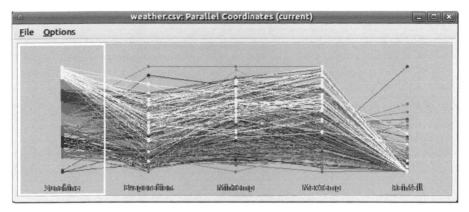

Figure 6.11: GGobi's parallel coordinates plot.

As with many of the approaches to data visualisation, when there are many observations the plots can become rather crowded and lose some of their usefulness. For example, a scatter plot over very many points will sometimes become a solid block of points showing little useful information.

Quality Plots Using R

We can save the plots generated by GGobi into an R script file and then have R generate the plots for us. This allows the plots to be regenerated as publication-quality graphics using R's capabilities. **DescribeDisplay** (Wickham et al., 2010) is required for this:

```
> install.packages("DescribeDisplay")
> library(DescribeDisplay)
```

Then, within GGobi, we choose from the Tools menu to Save Display Description. This will prompt us for a filename into which GGobi will write an R script to recreate the current graphic. We can load this script into R with `dd_load()` and then generate a plot in the usual way:

```
> pd <- dd_load("ggobi-saved-display-description.R")
> pdf("ggobi-rplot-deductions-outliers")
> plot(pd)
> dev.off()
> ggplot(pd)
```

R code can also be included in LibreOffice documents to directly generate and include the plots within the document using **odfWeave** (Kuhn et al., 2010). For Microsoft Word, **SWordInstaller** offers similar functionality.

Further GGobi Documentation

We have only really just started to scratch the surface of using GGobi here. There is a lot more functionality available, and whilst the functionality that is likely to be useful for the data miner has been touched on, there is a lot more to explore. So do explore the other features of GGobi, as some will surely be useful for new tasks. A very good overview of using GGobi for visual data mining is presented by Cook and Swayne (2007). Another overview is provided by Wickham et al. (2008).

6.3   Command Summary

This chapter has referenced the following R packages, commands, functions, and datasets:

`dd_load()`	command	Load an **rggobi** plot script file.
`dev.off()`	command	Close a graphics device.
`display()`	command	Create a new GGobi display.
`ggplot()`	command	Advanced plotting functionality.
`ggobi()`	command	Interactive data exploration using GGobi.
`latticist()`	command	Interactive data exploration within R.
latticist	package	Interactive data exploration within R.
odfWeave	package	Embed R in LibreOffice documents.
`plot()`	command	Visualise supplied data.
rggobi	package	Interactive data exploration using GGobi.
weather	dataset	Sample dataset from **rattle**.

Chapter 7

Transforming Data

An interesting issue with the delivery of a data mining project is that in reality we spend more of our time working on and with the data than we do building actual models, as we suggested in Chapter 1. In building models, we will often be looking to improve their performance. The answer is often to improve our data. This might entail sourcing some additional data, cleaning up the data, dealing with missing values in the data, transforming the data, and analysing the data to raise its efficiency through a better choice of variables.

In general, we need to transform our data from the raw data originally supplied for a data mining project to the polished and focussed data from which we build our best models. This is often the make-or-break phase of a data mining project.

This chapter introduces these data issues. We then review the various options for dealing with some of these issues, illustrating how to do so in Rattle and R.

7.1 Data Issues

A review of the winning entries in the annual data mining competitions reinforces the notion that building models from the right data is crucial to the success of a data mining project. The ACM KDD Cup, an annual Data Mining and Knowledge Discovery competition, is often won by a team that has placed a lot of effort in preprocessing the data supplied.

The 2009 ACM KDD Cup competition is a prime example. The French telecommunications company Orange supplied data related to customer relationship management. It consisted of 50,000 observations with much missing data. Each observation recorded values for 15,000 (anonymous) variables. There were three target variables to be modelled. One of the common characteristics for many entries was the preprocessing performed on the data. This included dealing with missing values, recoding data in various ways, and selecting variables. Some of the resulting models, for example, used only one or two hundred of the original 15,000 variables.

We review in this section some of the issues that relate to the quality of the data that we might have available for data mining. We then consider how we deal with these issues in the following sections.

An important point to understand is that often in data mining we are making use of, and indeed making do with, the data that is available. Such data might be regularly collected for other purposes. Some variables might be critical to the operation of the business and so special attention is paid to ensuring its accuracy. However, other data might only be informational and so less attention is paid to its quality.

We need to understand many different aspects about how and why the data was collected in order to understand any data issues. It is crucial to spend time understanding such data issues. We should do this before we start building models and then again when we are trying to understand why particular models have emerged. We need to explore the data issues that may have led to specific patterns or anomalies in our models. We may then need to rectify those issues and rebuild our models.

Data Cleaning

When collecting data, it is not possible to ensure it is perfectly collected, except in trivial cases. There will always be errors in the collection, despite how carefully it might have been collected. It cannot be stressed enough that we always need to be questioning the quality of the data we have. Particularly in large data warehouse environments where a lot of effort has already been expended in addressing data quality issues, there will still remain dirty data. It is important to always question the data quality and to be alert to the issue.

There are many reasons for the data to be dirty. Simple data entry errors occur frequently. Decimal points can be incorrectly placed, turning $150.00 into $15000. There can be inherent error in any counting or measuring device. There can also be external factors that cause errors to change over time, and so on.

One of the most important ongoing tasks we have in data mining, then, is cleaning our data. We usually start cleaning the data before we build our models. Exploring the data and building descriptive and predictive models will lead us to question the quality of the data at different times, particularly when we identify odd patterns.

A number of simple steps are available in reviewing the quality of our data. In exploring data, we will often explore variables through frequency counts and histograms. Any anomalous patterns there should be explored and explained. For categoric variables, for example, we would be on the lookout for categories with very low frequency counts. These might be mistyped or differently typed (upper/lowercase) categories.

A major task in data cleaning is often focussed around cleaning up names and addresses. This becomes particularly significant when bring ing data together from multiple sources. In combining financial and business data from numerous government agencies and public sources, for example, it is not uncommon to see an individual have his or her name recorded in multiple ways. Up to 20 or 30 variations can be possible. Street addresses present the same issues. A significant amount of effort is often expended in dealing with cleaning up such data in many organisations, and a number of tools have been developed to assist in the task.

Missing Data

Missing data is a common feature of any dataset. Sometimes there is no information available to populate some value. Sometimes the data has simply been lost, or the data is purposefully missing because it does not apply to a particular observation. For whatever reason the data is missing, we need to understand and possibly deal with it.

Missing values can be difficult to deal with. Often we will see missing values replaced with sentinels to mark that they are missing. Such sentinels can include things like 9999, or 1 Jan 1900, or even special characters that can interfere with automated processing like "*", "?", "#", or "$". We consider dealing with missing values through various transformations, as discussed in Section 7.4.

Outliers

An outlier is an observation that has values for the variables that are quite different from most other observations. Typically, an outlier appears at the maximum or minimum end of a variable and is so large or small that it skews or otherwise distorts the distribution. It is not uncommon to have a single instance or a very small number of these outlier values when compared to the frequency of other values of the variable. When summarising our data, performing tests on the data, and in building models, outliers can have an adverse impact on the quality of the results.

Hawkins (1980) captures the concept of an outlier as

> an observation that deviates so much from other observations as to arouse suspicion that it was generated by a different mechanism.

Outliers can be thought of as exceptional cases. Examples might include extreme weather conditions on a particular day, a very wealthy person who financially is very different from the rest of the population, and so on. Often, an outlier may be interesting but not really a key observation for our analysis. Sometimes outliers are the rare events that we are specifically interested in.

We may be interested in rare, unusual, or just infrequent events in a data mining context when considering fraud in income tax, insurance, and on-line banking, as well as for marketing.

Identifying whether an observation is an outlier is quite difficult, as it depends on the context and the model to be built. Perhaps under one context an observation is an outlier but under another context it might be a typical observation. The decision of what an outlier is will also vary by application and by user.

General outlier detection algorithms include those that are based on distance, density, projections, or distributions. The distance-based approaches are common in data mining, where an outlier is identified based on an observation's distance from nearby observations. The number of nearby observations and the minimum distance are two parameters. Another common approach is to assume a known distribution for the data. We then consider by how much an observation deviates from the distribution.

Many more recent model builders (including random forests and support vector machines) are very robust to outliers in that outliers tend not to adversely affect the algorithm. Linear regression type approaches tend to be affected by outliers.

One approach to dealing with outliers is to remove them from the dataset altogether. However, identifying the outlier remains an issue.

Variable Selection

Variable selection is another approach that can result in improved modelling. By removing irrelevant variables from the modelling process, the resulting models can be made more robust. Of course, it takes a good knowledge of the dataset and an understanding of the relevance of variables to the problem at hand. Some variables will also be found to be quite related to other variables, creating unnecessary noise when building models.

Various techniques can be used for variable selection. Simple techniques include considering different subsets of variables to explore for a subset that provides the best results. Other approaches use modelling measures (such as the information measure of decision tree induction discussed in Chapter 11) to identify the more important collection of variables.

A variety of other techniques are available. Approaches like principal components analysis and the variable importance measures of random forests and boosting can guide the choice of variables for building models.

7.2   Transforming Data

With the plethora of issues that we find in data, there is quite a collection of approaches for transforming data to improve our ability to discover knowledge. Cleaning our dataset and creating new variables from other variables in the dataset occupies much of our time as data miners. A programming language like R provides support for most of the myriad of approaches possible.

Rattle's Transform tab (Figure 7.1) provides many options for transforming datasets using many of the more common transformations. This includes normalising our data, filling in missing values, turning numeric variables into categoric variables and vice versa, dealing with outliers, and removing variables or observations with missing values. For the more complex transformations, we can revert to using R.

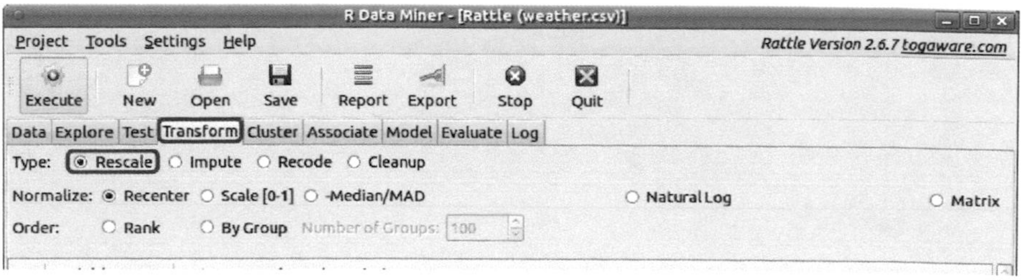

Figure 7.1: The Transform tab options.

We now introduce the various transformations supported by Rattle. In tuning our dataset, we will often transform it in many different ways. This often represents quite a lot of work, and we need to capture the resulting data in some form.

Once the dataset is transformed, we can save the new version to a CSV file. We do this by clicking on the Export button whilst viewing the Transform (or the Data) tab. This will prompt us for a CSV filename under which the current transformed dataset will be saved. We can also save the whole current state of Rattle as a project, which can easily be reloaded at a later time.

Another option, and one to be encouraged as good practise, is to save to a script file the series of transformations as recorded in the Log tab. Saving these to a script file means we can automate the generation of the transformed dataset from the original dataset. The automatically transformed dataset can then be used for building models or for scoring. For scoring (i.e., applying a model to a new collection of data), we can simply change the name of the original source data file within the script. The data is then processed through the R script and we can then apply our model to this new dataset within R.

The remainder of this chapter introduces each of the classes of transformations that are typical of a data mining project and supported by Rattle.

7.3   Rescaling Data

Different model builders will have different assumptions on the data from which the models are built. When building a cluster using any kind of distance measure, for example, we may need to ensure all variables have approximately the same scale. Otherwise, a variable like Income will overwhelm a variable like Age when calculating distances. A distance of 10 "years" may be more significant than a distance of \$10,000, yet 10000 swamps 10 when they are added together, as would be the case when calculating distances without rescaling the data.

In these situations, we will want to normalise our data. The types of normalisations (available through the Normalise option of the Transform tab) we can perform include recentering and rescaling our data to be around zero (Recenter uses a so-called Z score, which subtracts the mean and divides by the standard deviation), rescaling our data to be in the range from 0 to 1 (Scale [0–1]), performing a robust rescaling around zero using the median (Median/MAD), applying log() to our data, or transforming multiple variables with one divisor (Matrix). The details of these transformations will be presented below.

Other rescaling transformations include converting the numbers into a rank ordering (Rank) and performing a transform to rescale a variable according to some group that the observation belongs to (By Group).

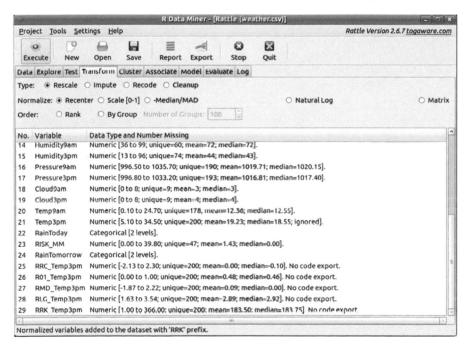

Figure 7.2: Transforming Temp3pm in five different ways.

Figure 7.2 shows the result of transforming the variable Temp3pm in five different ways. The simple summary that we can see for each variable in Figure 7.2 provides a quick view of how the data has been transformed. For example, the recenter transform of the variable `Temp3pm` has changed the range of values for the variable from the original 5.10 to 34.50 to end up with −2.13 to 2.30.

> **Tip:** *Notice, as we see in Figure 7.2, that the original data is not modified. Instead, a new variable is created for each transform with a prefix added to the variable's name that indicates the kind of transformation. The prefixes are RRC_ (for **Recenter**), RO1_ (for **Scale [0–1]**), RMD_ (for **Median/MAD**), RLG_ (for **Log**), and RRK_ (for **Rank**).*

Figure 7.3 illustrates the effect of the four transformations on the variable `Temp3pm` compared with the original distribution of the data. The top left plot shows the original distribution. Note that the three normalisations (recenter, rescale 0–1, and recenter using the median/-MAD) all produce new variables with very similar looking distributions. The log transform changes the distribution quite significantly. The rank transform simply delivers a variable with a flat distribution since the new variable simply consists of a sequence of integers and thus each value of the new variable appears just once.

Recenter

This is a common normalisation that re-centres and rescales our data. The usual approach is to subtract the mean value of a variable from each observation's value of the variable (to recentre the variable) and then divide the values by their standard deviation (calculating the square root of the sum of squares), which rescales the variable back to a range within a few integer values around zero.

To demonstrate the transforms on our *weather*, we will load **rattle** and create a copy of the dataset, to be referred to as `ds`:

```
> library(rattle)
> ds <- weather
```

The following R code can then perform the transformation using `scale()`:

```
> ds$RRC_Temp3pm <- scale(ds$Temp3pm)
```

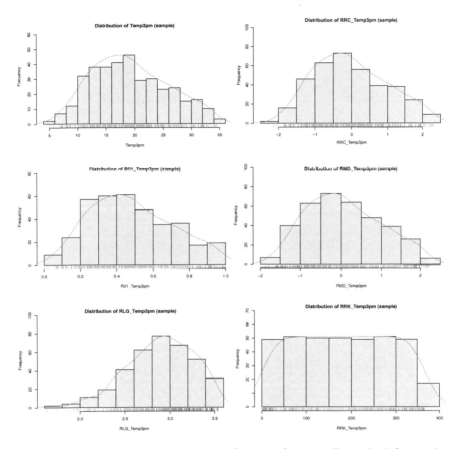

Figure 7.3: Comparing distributions after transforming. From the left to right, top to bottom: original, recenter, rescale to 0–1, rank, log transform, and recenter using median/MAD.

Scale [0–1]

Rescaling so that our data has a mean around zero might not be so intuitive for variables that are never negative. Most numeric variables from the *weather* dataset naturally only take on positive values, including `Rainfall` and `WindSpeed3pm`.

To rescale whilst retaining only positive values, we might choose the Scale [0–1] transform, which simply recodes the data so that the values are all between 0 and 1. This is done by subtracting the minimum value from the variable's value for each observation and then dividing by the difference between the minimum and the maximum values.

The following R code is used to perform the transformation. We use `rescaler()` from **reshape** (Wickham, 2007):

```
> library(reshape)
> ds$RO1_Temp3pm <- rescaler(ds$Temp3pm, "range")
```

Median/MAD

This option for recentring and rescaling our data is regarded as a robust (to outliers) version of the standard Recenter option. Instead of using the mean and standard deviation, we subtract the median and divide by the so-called median absolute deviation (MAD).

The following R code is used to perform the transformation. Again we use `rescaler()` from **reshape**:

```
> library(reshape)
> ds$RMD_Temp3pm <- rescaler(ds$Temp3pm, "robust")
```

Natural Log

Often the values of a variable can be quite skewed in one direction or another. A typical example is Income. The majority of a population may have incomes below $150,000. But there are a relatively small number of individuals with excessive incomes measured in the millions of dollars. In many approaches to analysis and model building, these extreme values (outliers) can adversely affect any analysis.

Logarithm transforms map a very broad range of (positive) numeric values into a narrower range of (positive) numeric values. The natural log function effectively reduces the spread of the values of the variable. This is particularly useful when we have outliers with extremely large values compared with the rest of the population.

Logarithms can use a so called *base* with respect to which they do the transformation. We can use a base 10 transform to explain what the transform does. With a $\log_{10}$ transform, a salary of $10,000 is recoded as 4, $100,000 as 5, $150,000 as 5.17609125905568, and $1,000,000 as 6—that is, a logarithm of base 10 recodes each power of 10 (e.g., 10^5 or 100,000) to the power itself (e.g., 5) and similarly for a logarithm of base 2, which recodes 8 (which is 2^3) to 3.

By default, Rattle simply uses the natural logarithm for its transform. This recodes using a logarithm to base e, where e is the special number 2.718.... This is the default base that R uses for log(). The following R code is used to perform the transformation. We also recode any resulting "infinite" values (e.g., $\log(0)$) to be treated as missing values:

```
> ds$RLG_Temp3pm <- log(ds$Temp3pm)
> ds$RLG_Temp3pm[ds$RLG_Temp3pm == -Inf] <- NA
```

Rank

On some occasions, we are not interested in the actual value of the variable but rather in the relative position of the value within the distribution of that variable. For example, in comparing restaurants or universities, the actual score may be less interesting than where each restaurant or university sits compared with the others. A rank is then used to capture the relative position, ignoring the actual scale of any differences.

The Rank option will convert each observation's numeric value for the identified variable into a ranking in relation to all other observations in the dataset. A rank is simply a list of integers, starting from 1, that is mapped from the minimum value of the variable, progressing by integer until we reach the maximum value of the variable. The largest value is thus the sample size, which for the *weather* dataset is 366. A rank has an advantage over a recentring transform, as it removes any skewness from the data (which may or may not be appropriate for the task at hand).

A problem with recoding our data using a rank is that it becomes difficult when using the resulting model to score new observations. How do we rank a single observation? For example, suppose we have a model that tests whether the rank is less than 50 for the variable Temp3pm. What does this actually mean when we apply this test to a new observation? We might instead need to revert the rank back to an actual value to be useful in scoring.

The following R code is used to perform the transformation. Once again we use **rescaler()** from **reshape**:

```
> library(reshape)
> ds$RRK_Temp3pm <- rescaler(ds$Temp3pm, "rank")
```

By Group

A By Group transform recodes the values of a variable into a rank order between 0 and 100. A categoric variable can also be identified as part of the transformation. In this case, the observations are grouped by the values of the categoric variable. These groups are then considered as peers. The ranking is then performed with respect to the peers rather than the whole population. An example might be to rank wind speeds within groups defined by the wind direction. A high wind speed relative to one direction may not be a high wind speed relative to another direction. The code to do this gets a little complex.

```
> library(reshape)
> ds$RBG_SpeedByDir <- ds$WindGustSpeed
> bylevels <- levels(ds$WindGustDir)
> for (vl in bylevels)
  {
    grp <- sapply(ds$WindGustDir == vl, isTRUE)
    ds[grp, "RBG_SpeedByDir"] <-
      round(rescaler(ds[grp, "WindGustSpeed"],
                     "range") * 99)
  }
> ds[is.nan(ds$RBG_SpeedByDir), "RBG_SpeedByDir"] <- 50
> v <- c("WindGustSpeed", "WindGustDir", "RBG_SpeedByDir")
```

We can then selectively display some observations:

```
> head(ds[ds$WindGustDir %in% c("NW", "SE"), v], 10)
```

Observation 1, for example, with a WindGustSpeed of 30, is at the 18th percentile within all those observations for which WindGustDir is NW. Overall, we might observe that the WindGustSpeed is generally less when the WindGustDir is SE as compared with NW, looking at the rankings within each group. Instead of generating a rank of between 0 and 100, a Z score (i.e., Recenter) could be used to recode within each group. This would require only a minor change to the R code above.

Summary

We summarise this collection of transformations of the first few observations of the variable Temp3pm:

Obs.	WindGustSpeed	WindGustDir	RBG_SpeedByDir
1	30	NW	18
3	85	NW	83
4	54	NW	47
6	44	SE	86
7	43	SE	83
14	44	NW	35
15	41	NW	31
29	39	SE	70
33	50	NW	42
34	50	NW	42

Obs.	Temp3pm	RRC_	R01_	RMD_	RLG_	RRK_
1	23.60	0.66	0.63	0.70	3.16	268
2	25.70	0.97	0.70	0.99	3.25	295
3	20.20	0.15	0.51	0.23	3.01	217
4	14.10	−0.77	0.31	−0.62	2.65	92
5	15.40	−0.58	0.35	−0.44	2.73	117
6	14.80	−0.67	0.33	−0.52	2.69	106

7.4 Imputation

Imputation is the process of filling in the gaps (or missing values) in data. Data is missing for many different reasons, and it is important to understand why. This will guide us in dealing with the missing values. For rainfall variables, for example, a missing value may mean there was no rain recorded on that day, and hence it is really a surrogate for 0 mm of rain. Alternatively, perhaps the measuring equipment was not functioning that day and hence recorded no rain.

Imputation can be questionable because, after all, we are inventing data. We won't discuss here the pros and cons in any detail, but note that, despite such concerns, reasonable results can be obtained from simple imputations.

There are many types of imputations available, only some of which are directly available in Rattle. Imputation might involve simply replacing missing values with a particular value. This then allows, for example, linear regression models to be built using all observations. Or we might add an additional variable to record when values are missing. This then allows the model builder to identify the importance of the missing values, for example. We do note, however, that not all model builders (e.g., decision trees) are troubled by missing values.

Figure 7.4 shows Rattle's Impute option on the Transform tab selected with the choices for imputation, including Zero/Missing, Mean, Median, Mode, and Constant.

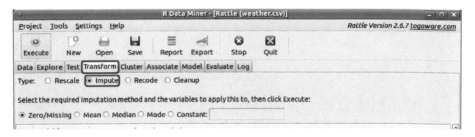

Figure 7.4: The Transform tab with the Impute option selected.

When Rattle performs an imputation, it will store the results in a new variable within the same dataset. The new variable will have the same name as the variable that is imputed, but prefixed with either IZR_, IMN_, IMD_, IMO_, or ICN_. Such variables will automatically be identified as having an Input role, whilst the original variable will have a role of Ignore.

Zero/Missing

The simplest imputations involve replacing all missing values for a variable with a single value. This makes the most sense when we know that the missing values actually indicate that the value is 0 rather than unknown. For example, in a taxation context, if a taxpayer does not provide a value for a specific type of deduction, then we might assume that they intend it to be zero. Similarly, if the number of children in a family is not recorded, it could be a reasonable assumption to assume it is zero.

For categoric data, the simplest approach to imputation is to replace missing values with a special value, such as *Missing*. The following R code is used to perform the transformation:

```
> ds$IZR_Sunshine <- ds$Sunshine
> ds$IZR_Sunshine[is.na(ds$IZR_Sunshine)] <- 0
```

Mean/Median/Mode

Often a simple, if not always satisfactory, choice for missing values that are known not to be zero is to use some "central" value of the variable. This is often the mean, median, or mode, and thus usually has limited impact on the distribution. We might choose to use the mean, for example, if the variable is otherwise generally normally distributed (and in particular does not have any skewness). If the data does exhibit some skewness, though (e.g., there are a small number of very large values), then the median might be a better choice.

For categoric variables, there is, of course, no mean or median, and so in such cases we might (but with care) choose to use the mode (the most frequent value) as the default to fill in for the otherwise missing values. The mode can also be used for numeric variables. This could be appropriate for variables that are dominated by a single value. Perhaps we notice that predominately (e.g., for 80% of the observations) the temperature at 9 am is 26 degrees Celsius. That could be a reasonable choice for any missing values.

Whilst this is a simple and computationally quick approach, it is a very blunt approach to imputation and can lead to poor performance from the resulting models. However, it has also been found empirically to be useful. The following R code is used to perform the transformation:

```
> ds$IMN_Sunshine <- ds$Sunshine
> ds$IMN_Sunshine[is.na(ds$IMN_Sunshine)] <-
    mean(ds$Sunshine, na.rm=TRUE)
```

Constant

This choice allows us to provide our own default value to fill in the gaps. This might be an integer or real number for numeric variables, or else a special marker or the choice of something other than the majority category for categoric variables. The following R code is used to perform the transformation:

```
> ds$IZR_Sunshine <- ds$Sunshine
> ds$IZR_Sunshine[is.na(ds$IZR_Sunshine)] <- 0
```

7.5   Recoding

The Recode option on the Transform tab provides numerous remapping operations, including binning and transformations of the type of the data. Figure 7.5 lists the options.

Figure 7.5: The Transform tab with the Recode option selected.

Binning

Binning is the operation of transforming a continuous numeric variable into a specific set of categoric values based on the numeric values. Simple examples include converting an age into an age group, and a temperature into Low, Medium, and High. Performing a binning transform may lose valuable information, so do give some thought as to whether binning is appropriate.

Binning can be useful in simplifying models. It is also useful when we visualise data. A mosaic plot (Chapter 5), for example, is only useful for categoric data, and so we could turn Sunshine into a categoric variable by binning. Binning can also be useful to set a numeric value as the stratifying variable in various plots in Chapter 5. For example, we could bin Temp9am and then choose the new BE4_Temp9am (BE4 for binning into four equal-size bins) as the Target and generate a Box Plot from the Explore tab to see the relationship with the Evaporation.

Rattle supports automated binning through the use of binning() (provided by Daniele Medri). The Rattle interface provides an option to choose between Quantile (or equal count) binning, KMeans binning, and Equal Width binning. For each option, the default number of bins is four. We can change this to suit our needs. The variables generated are prefixed with either BQn_, BKn_, or BEn_, respectively, with n replaced by the number of bins.

Indicator Variables

Some model builders often do not directly handle categoric variables. This is typical of distance-based model builders such as k-means clustering, as well as the traditional numeric regression types of models.

A simple approach to transforming a categoric variable into a numeric one is to construct a collection of so-called *indicator* or *dummy* variables. For each possible value of the categoric variable, we can create a new variable that will have the value 1 for any observation that has this categoric value and 0 otherwise. The result is a collection of new numeric variables, one for each of the possible categoric values. An example might be the categoric variable Colour, which might only allow the possible values of Red, Green, or Blue. This can be converted to three variables, Colour_Red, Colour_Green, and Colour_Blue. Only one of these will have the value 1 at any time, whilst the other(s) will have the value 0.

Rattle's Transform tab provides an option to transform a categoric variable into a collection of indicator variables. Each of the new variables has a name that is prefixed by TIN_. The remainder of the name is made up of the original name of the categoric variable (e.g., Colour) and the particular value (e.g., Red). This will give, for example, TIN_Colour_Red as one of the new variable names. Table 7.1 illustrates how the recoding works for a collection of observations.

Table 7.1: Examples of recoding a single categoric variable as a number of numeric indicator variables.

Obs.	Colour	Colour_Red	Colour_Green	Colour_Blue
1	Green	0	1	0
2	Blue	0	0	1
3	Blue	0	0	1
4	Red	1	0	0
5	Green	0	1	0
6	Red	1	0	0

In terms of modelling, for a categoric variable with k possible values, we only need to convert it to $k-1$ indicator variables. The kth indicator variable is redundant and in fact is directly determined by the values of the other $k-1$ indicators. If all of the other indicators are 0, then clearly the kth will be 1. Similarly if any of the other $k-1$ indicators is 1, then the kth must be 0. Consequently, we should only include all but one of the new indicator variables as having an Input role. Rattle, by default, will set the role of the first new indicator variable to be Ignore.

There is not always a need to transform a categoric variable. Some model builders, like the Linear model builder in Rattle, will do it automatically.

Join Categorics

The Join Categorics option provides a convenient way to stratify the dataset based on multiple categoric variables. It is a simple mechanism that creates a new variable from the combination of all of the values of the two constituent variables selected in the Rattle interface. The resulting variables are prefixed with TJN_ and include the names of both the constituent variables.

A simple example might be to join RainToday and RainTomorrow to give a new variable (TJN here and TJN_RainToday_RainTomorrow in Rattle):

```
> ds$TJN <- interaction(paste(ds$RainToday, "_",
                              ds$RainTomorrow, sep=""))
> ds$TJN[grepl("^NA_|_NA$", ds$TJN)] <- NA
> ds$TJN <- as.factor(as.character(ds$TJN))
> head(ds[c("RainToday", "RainTomorrow", "TJN")])

  RainToday RainTomorrow     TJN
1        No          Yes  No_Yes
2       Yes          Yes Yes_Yes
3       Yes          Yes Yes_Yes
4       Yes          Yes Yes_Yes
5       Yes           No  Yes_No
6        No           No   No_No
```

We might also want to join a numeric variable and a categoric variable, like the common Age and Gender stratification. To do this, we first use the Binning option within Recode to categorise the Age variable and then use Join Categorics.

Type Conversion

The As Categoric and As Numeric options will, respectively, convert a numeric variable to categoric (with the new categoric variable name prefixed with TFC_) and vice versa (with the new numeric variable name prefixed with TNM_). The R code for these transforms uses `as.factor()` and `as.numeric()`:

```
> ds$TFC_Cloud3pm <- as.factor(ds$Cloud3pm)
> ds$TNM_RainToday <- as.numeric(ds$RainToday)
```

7.6 Cleanup

It is quite easy to get our dataset variable count up to significant numbers. The Cleanup option allows us to tell Rattle to actually delete columns from the dataset. Thus, we can perform numerous transformations and then save the dataset back into a CSV file (using the Export option).

Various Cleanup options are available. These allow us to remove any variable that is ignored (Delete Ignored), remove any variables we select (Delete Selected), or remove any variables that have missing values (Delete Missing). The Delete Obs with Missing option will remove observations (rather than variables—i.e., remove rows rather than columns) that have missing values.

7.7 Command Summary

This chapter has referenced the following R packages, commands, functions, and datasets:

`as.factor()`	function	Convert variable to be categoric.
`as.numeric()`	function	Convert variable to be numeric.
`is.na()`	function	Identify which values are missing.
`levels()`	function	List the values of a categoric variable.
`log()`	function	Logarithm of a numeric variable.
`mean()`	function	Mean value of a numeric variable.
`rescaler()`	function	Remap numeric variables.

reshape   package   Transform variables in various ways.
`scale()`   function   Remap numeric variables.

Part II

Building Models

Chapter 8

Descriptive and Predictive Analytics

Modelling is what we most often think of when we think of data mining. Modelling is the process of taking some data (usually) and building a simplified description of the processes that might have generated it. The description is often a computer program or mathematical formula. A model captures the knowledge exhibited by the data and encodes it in some language. Often the aim is to address a specific problem through modelling the world in some form and then use the model to develop a better understanding of the world.

We now turn our attention to building models. As in any data mining project, building models is usually the aim, yet we spend a lot more time understanding the business problem and the data, and working the data into shape, before we can begin building the models. Often we gain much valuable knowledge from our preparation for modelling, and some data mining projects finish at that stage, even without the need to build a model—that might be unusual, though, and we do need to expect to build a model or two. As we will find, we build models early on in a project, then work on our data some more to transform, shape, and clean it, build more models, then return to processing the data once again, and so on for many iterations. Each cycle takes us a step closer to achieving our desired outcomes.

This chapter introduces the concept of models and model builders that fall into the categories of data mining: descriptive and predictive. In this chapter, we provide an overview of these approaches. For descriptive data mining, we present cluster analysis and association rules as two approaches to model building. For predictive data mining, we consider both classification and regression models, introducing algorithms like decision trees, random forests, boosting, support vector machines, linear regression, and neural networks. In each case, in their own chapters, the algorithms are presented together with a guide to using them within Rattle and R.

8.1   Model Nomenclature

Much of the terminology used in data mining has grown out of terminology used in both machine learning and research statistics. We identify, for example, two very broad categories of model building algorithms as **descriptive analytics** and **predictive analytics**. In a traditional machine learning context, these equate to **unsupervised learning** and **supervised learning**. We cover both approaches in the following chapters and describe each in a little more detail in the following sections.

On top of the basic algorithm for building models, we also identify **meta learners**, which include **ensemble learners**. These approaches suggest building many models and combining them in some way. Some ideas for ensembles originate from the multiple inductive learning (MIL) algorithm (Williams, 1988), where multiple decision tree models are built and combined as a single model.

8.2   A Framework for Modelling

Building models is a common pursuit throughout life. When we think about it, we build ad hoc and informal models every day when we solve problems in our head and live our lives. Different professions, like architects and engineers, for example, specifically build models to see how things fit together, to make sure they do fit together, to see how things will work in the real world, and even to sell the idea behind the model to others. Data mining is about building models that give us insights into the world and how it works. But even more than that, our models are often useful to give us guidance in how to deal with and interact with the real world.

Building models is thus fundamental to understanding our world. We start doing it as a child and continue until death. When we build a model, whether it be with toy bricks, papier mâché, or computer software, we get a new perspective of how things fit together or interact. Once we have some basic models, we can start to get ideas about more complex ones, building on what has come before. With data mining, our models are driven by the data and thus aim to be objective. Other models might be more subjective and reflect our views of what we are modelling.

In understanding new, complex ideas, we often begin by trying to map the idea into concepts or constructs that we already know. We bring these constructs together in different ways that reflect how we understand a new, more complex idea. As we learn more about the new, complex idea, we change our model to better reflect that idea until eventually we have a model that is a good enough match to the idea.

The same is true when building models using computers. Writing any computer program is essentially about building a model. An accountant's spreadsheet is a model of something in the world. A social media application captures a model or introduces a new model of how people communicate. Models of the economy and of the environment provide insights into how these things work and allow us to explore possible future scenarios.

An important thing to remember, though, is that no model can perfectly represent the real world, except in the most simplistic and trivial of scenarios. To perfectly model the real world, even if it were possible, we would need to incorporate into the model every possible variable imaginable. The real world has so many different factors feeding into it that all we can really hope to do is to get a good approximation of it.

A model, as a good approximation of the world, will express some understanding of that world. It needs to be expressed using some language, whether it be a spoken or written human language, a mathematical language, a computer language, or a modelling language. The **language** is used to represent our knowledge.

We write or speak in **sentences** based on the language we have chosen. Some **sentences** expressed in our chosen language will capture useful knowledge. Other sentences might capture misinformation, and yet others may capture beliefs or propositions, and so on. Formally, each sentence will express or capture some concept within the formal constraints of the particular language chosen. We can think of constructing a sentence to express something about our data as building a model.

For any language, though, there is often an infinite (or at least a very large) collection of possible sentences (i.e., models) that can be expressed. We need some way of **measuring** how good a sentence is. This might just be a measure of how well formed our written sentence is— is it grammatically correct and does it read well? But just as importantly, does the sentence express a valid statement about the world? Does it provide useful insight and knowledge about the world? Is it a good model?

For each of the model builders we introduce, we will use this three-pronged framework:

- identify the **language** used to express the discovered knowledge,

- develop a mechanism to **search** for good sentences within the language, and

- define a **measure** that can be used to assess how good a sentence is.

This is a quite common framework from the artificial intelligence tradition. There we seek to automatically search for solutions to problems, within the bounds of a chosen knowledge representation language.

This framework is simply cast for the task of data mining—the task of building models. We refer to an algorithm for building a model as a **model builder**. Rattle supports a number of model builders, including clustering, association rules, decision tree induction, random forests, boosted decision trees, support vector machines, logistic regression, and neural networks. In essence, the model builders differ in how they represent the models they build (i.e., the discovered knowledge) and how they find (or search for) the best model within this representation.

In building a model, we will often look to the structure of the model itself to provide insights. In particular, we can learn much about the relationships between the input variables and the target variable (if any) from studying our models. Sometimes these observations themselves deliver benefits from the data mining project, even without actually using the models directly.

There is generally an infinite number of possible sentences (i.e., models) given any specific language. In human language, we are generally very well skilled at choosing sentences from this infinite number of possibilities to best represent what we would like to communicate. And so it needs to be with model building. The skill is to express, within the chosen language, the best sentences that capture what it is we are attempting to model.

8.3   Descriptive Analytics

Descriptive analytics is the task of providing a representation of the knowledge discovered without necessarily modelling a specific outcome. The tasks of cluster analysis, association and correlation analysis and pattern discovery, can fall under this category.

From a machine learning perspective, we might compare these algorithms to unsupervised learning. The aim of unsupervised learning is to identify patterns in the data that extend our knowledge and understanding of the world that the data reflects. There is generally no specific target variable that we are attempting to model. Instead, these approaches shed light on the patterns that emerge from the descriptive analytics.

8.4   Predictive Analytics

Often our task in data mining is to build a model that can be used to predict the occurrence of an event. The model builders will extract knowledge from historic data and represent it in such a form that we can apply the resulting model to new situations. We refer to this as predictive analytics.

The tasks of classification and regression are at the heart of what we often think of as data mining and specifically predictive analytics. Indeed, we call much of what we do in data mining predictive analytics.

From a machine learning perspective, this is also referred to as supervised learning. The historic data from which we build our models will already have associated with it specific outcomes. For example, each observation of the *weather* dataset has associated with it a known outcome, recorded as the target variable. The target variable is `RainTomorrow` (whether it rained the following day), with the possible values of `No` and `Yes`.

Classification models are used to predict the class of new observations. New observations are classified into the different target variable categories or classes (for the *weather* dataset, this would be Yes and No). Often we will be presented with just two classes, but it could be more. A new observation might be today's weather observation. We want to classify the observation into the class Yes or the class No. Membership in a particular class indicates whether there might be rain on the following day or not, as the case may be.

Often, classification models are represented symbolically. That is, they are often expressed as, for example, a series of tests (or conditions) on different variables. Each test exhibits a piece of the knowledge that, together with other tests, leads to the identified outcome.

Regression models, on the other hand, are generally models that predict a numeric outcome. For the *weather* dataset, this might be the amount of rain expected on the following day rather than whether it will or won't rain. Regression models are often expressed as a mathematical formula that captures the relationship between a collection of input variables and the numeric target variable. This formula can then be applied to new observations to predict a numeric outcome.

Interestingly, *regression* comes from the word "regress," which means to move backwards. It was used by Galton (1885) in the context of techniques for regressing (i.e., moving from) observations to the average. The early research included investigations that separated people into different classes based on their characteristics. The regression came from modelling the heights of related people (Crano and Brewer, 2002).

8.5   Model Builders

Each of the following chapters describes a particular class of model builders using specific algorithms. For each model builder, we identify the structure of the language used to describe a model. The search algorithm is described as well as any measures used to assist in the search and to identify a good model.

Following the formal overview of each model builder, we then describe how the algorithm is used in Rattle and R and provide illustrative examples. The aim is to provide insight into how the algorithm works and some details related to it so that as a data miner we can make effective use of the model builder.

The algorithms we present will generally be in the context of a two-class classification task where appropriate. The aim of such tasks is to distinguish between two classes of observations. Such problems abound. The two classes might, for example, identify whether or not it is predicted to rain tomorrow (No and Yes). Or they might distinguish between high-risk and low-risk insurance clients, productive and unproductive taxation audits, responsive and nonresponsive customers, successful and unsuccessful security breaches, and so on. Many of the popular algorithms are covered in the following chapters. Algorithms not covered include neural networks, linear and logistic regressions, and Bayesian approaches.

In demonstrating the tasks using Rattle (together with a guide to the underlying R code), we note that Rattle presents a basic collection of tuning parameters. Good default values for various options allow the user to more simply build a model with little tuning. However, this may not always be the right approach, and whilst it is certainly a good place to start, experienced users will want to make much more use of the fuller set of tuning parameters available directly through the R Console.

Chapter 9

Cluster Analysis

The clustering technique is one of the core tools that is used by the data miner. Clustering gives us the opportunity to group observations in a generally unguided fashion according to how similar they are. This is done on the basis of a measure of the distance between observations. For example, we might have a dataset that is made up of school children of various heights, a range of weights, and different 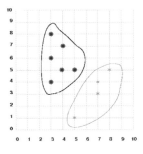 ages. Depending on what is needed to solve the problem at hand, we might wish to group the students into smaller, more definable groups and then compare different variables common to all groupings. Each group may have different ranges, minimums and maximums, and so on that represent that group. Clustering allows the data miner to break data into more meaningful groups and then contrast the different clusters against each other. Clusters can also be useful in grouping observations to help make the smaller datasets easier to manage. The aim of clustering is often to identify groups of observations that are close together but as a group are quite separate from other groups.

Numerous algorithms have been developed for clustering. In this chapter, we focus primarily on the k-means clustering algorithm. The algorithm will identify a collection of k clusters using a heuristic search starting with a selection of k randomly chosen clusters.

9.1   Knowledge Representation

A model built using the k-means algorithm represents the clusters as a collection of k means. The observations in the dataset are associated with their closest "mean" and thus are partitioned into k clusters. The *mean* of a particular *numeric* variable for a collection of observations is the average value of that variable over those observations. The *means* for the collection of observations that form one of the k clusters in any particular clustering are then the collection of mean values for each of the input variables over the observations within the clustering.

Consider, for example, a simple and small random subset of the *weather* dataset. This can be generated as below, where we choose only a small number of the available numeric variables:

```
> library(rattle)
> set.seed(42)
> obs1 <- sample(1:nrow(weather), 5)
> vars <- c("MinTemp", "MaxTemp",
            "Rainfall", "Evaporation")
> cluster1 <- weather[obs1, vars]
```

We now obtain the means of each of the variables. The vector of means then represents one of the clusters within our set of k clusters:

```
> mean(cluster1)
    MinTemp     MaxTemp    Rainfall Evaporation
       4.74       15.86        3.16        3.56
```

Another cluster will have a different mean:

```
> obs2 <- setdiff(sample(1:nrow(weather), 20), obs1)
> cluster2 <- weather[obs2, vars]
> mean(cluster2)
    MinTemp     MaxTemp    Rainfall Evaporation
     6.6474     19.7579      0.8421      4.4105
```

In comparing the two clusters, we might suggest that the second cluster generally has warmer days with less rainfall. However, without having actually built the clustering model, we can't really make too many such general observations without knowing the actual distribution of the observations.

A particular sentence in our knowledge representation language for k-means is then a collection of k sets of mean values for each of the variables. Thus, if we were to simply partition the *weather* dataset into ten sets (a common value for k), we would get ten sets of means for each of the four variables. Together, these 10 by 4 means represent a single sentence (or model) in the k-means "language."

9.2 Search Heuristic

For a given dataset, there are a very large number of possible k-means models that could be built. We might think to enumerate every possibility and then, using some measure that indicates how good the clustering is, choose the one that gets the best score. In general, this process of completely enumerating all possibilities would not be computationally possible. It may take hours, days, or weeks of computer time to generate and measure each possible set of clusters. Instead, the k-means algorithm uses a search heuristic. It begins with a random collection of k clusters. Each cluster is represented by a vector of the mean values for each of the variables.

The next step in the process is to then measure the distance between an observation and each of the k vectors of mean values. Each observation is then associated with its closest cluster.

We then recalculate the mean values based on the observations that are now associated with each cluster. This will provide us with a new collection of k vectors of means. With this new set of k means, we once again calculate the distance each observation is from each of the k means and reassociate the observation with the closest of the k means. This will often result in some observations moving from one group or cluster to another.

Once again, we recalculate the mean values based on the observations that are now associated with each cluster. Again, we have k new vectors of means. We repeat the process again. This iterative process is repeated until no more observations move from one cluster to another. The resulting clustering is then the model.

9.3   Measures

The basic measure used in building the model is a measure of distance, or conversely the measure of similarity between observations and the cluster means.

Any distance measure that measures the distance between two observations a and b must satisfy the following requirements:

- $d(a, b) \geq 0$ distance is nonnegative

- $d(a, a) = 0$ distance to itself is 0

- $d(a, b) = d(b, a)$ distance is symmetric

- $d(a, b) \leq d(a, c) + d(c, b)$ triangular inequality

One common distance measure is known as the Minkowski distance. This is formulated as

$$d(a, b) = \sqrt[q]{(|a_1 - b_1|^q + |a_2 - b_2|^q + \ldots + |a_n - b_n|^q)},$$

where a_1 is the value of variable 1 for observation a, etc. The value of q determines an actual distance formula. We can best picture the distance calculation using just two variables, like `MinTemp` and `MaxTemp`, from two observations. We plot the first two observations from the *weather* dataset in Figure 9.1 as generated using the following call to `plot()`. We also report the actual values being plotted.

```
> x <- round(weather$MinTemp[1:2])
> y <- round(weather$MaxTemp[1:2])
> plot(x, y, ylim=c(23, 29), pch=4, lwd=5,
        xlab="MinTemp", ylab="MaxTemp",
        bty="n")
> round(x)

[1]  8 14

> round(y)

[1] 24 27
```

When $q = 1$, d is known as the Manhattan distance:

$$d(a, b) = |a_1 - b_1| + |a_2 - b_2| + \ldots + |a_n - b_n|.$$

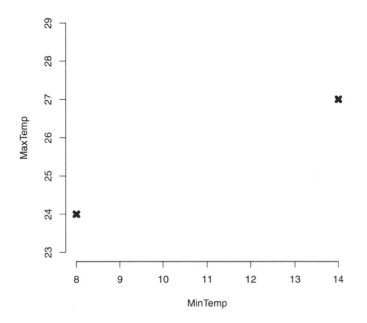

Figure 9.1: Two observations of two variables from the *weather* dataset. What are the possible ways of measuring the distance between these two points?

The Manhattan distance measure gets its name from one of the five boroughs of New York City. Most of the streets of Manhattan are laid out on a regular grid. Each block is essentially a rectangle. Figure 9.2 simplifies the grid structure but illustrates the point. Suppose we want to calculate the distance to walk from one block corner, say West 31st Street and 8th Avenue, to another, say West 17th Street and 6th Avenue. We must travel along the street, and the distance is given by how far we travel in each of just two directions, as is captured in the formula above.

For our *weather* dataset, we can add a grid() to the plot and limit our walk to the lines on the grid, as in Figure 9.2. The distance travelled will be $d = 6 + 3 = 9$, and one such path is shown as the horizontal and then vertical line in Figure 9.2.

When $q = 2$, d is known as the more familiar, and most commonly

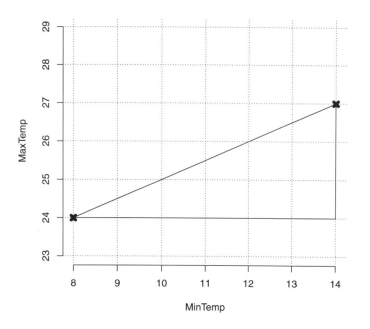

Figure 9.2: Measuring the distance by travelling the streets of Manhattan (the regular grid, with one path shown as the horizontal and then the vertical line), rather than as a bird might fly (the direct line between the two points).

used, Euclidean distance:

$$d(a, b) = \sqrt{(|a_1 - b_1|^2 + |a_2 - b_2|^2 + \ldots + |a_n - b_n|^2)}.$$

This is the straight-line distance between the two points shown in Figure 9.2. It is how a bird would fly direct from one point to another if it was flying high enough in Manhattan. The distance in this case is $d = \sqrt{6^2 + 3^2} = 6.32$.

In terms of how we measure the quality of the actual clustering model, there are very many possibilities. Most relate to measuring the distance between all of the observations within a cluster and summing that up. Then compare that with some measure of the distances between the means or even the observations of each of the different clusters. We will see and explain in a little more detail some of these measures in the next section.

9.4   Tutorial Example

The *weather* dataset is used to illustrate the building of a cluster model. The **Cluster** tab in the **Rattle** window provides access to various clustering algorithms, including k-means. `kmeans()` is provided directly through R by the standard **stats** package.

Building a Model Using Rattle

After loading a dataset into **Rattle**, we select the **Cluster** tab to be presented with various clustering algorithms. We will also see a simple collection of options available for use to fine-tune the model building. The k-means algorithm is the default option, and by default ten clusters will be built as the model. A random seed is provided. Changing the seed will result in a randomly different collection of starting points for our means. The heuristic search then begins the iterative process as described in Section 9.2.

Load the *weather* dataset from the **Data** tab, and then simply clicking the **Execute** button whilst on the **Cluster** tab will result in the k-means clustering output shown in Figure 9.3.

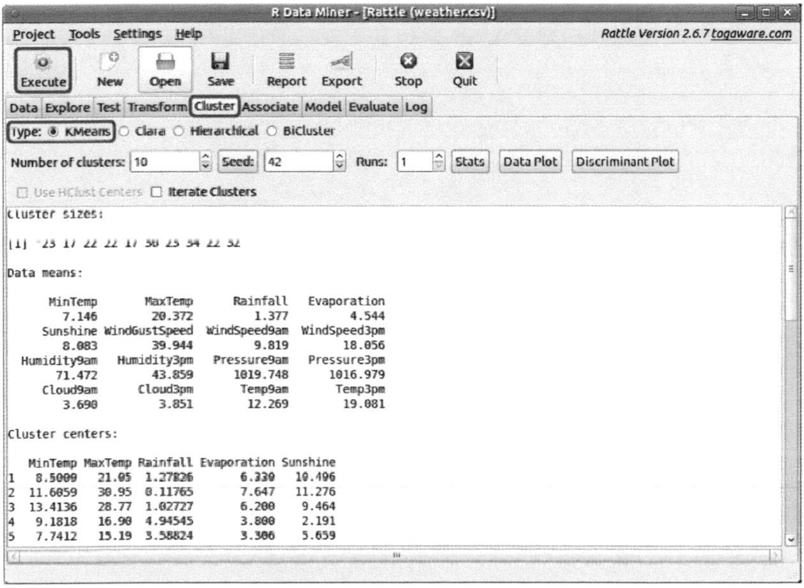

Figure 9.3: Building a k-means clustering model.

The text view contains a little information about the model that has been built. We will work our way through its contents. It begins with the cluster size, which is simply a count of the number of observations within each cluster:

```
Cluster sizes:


[1] "23 17 22 22 17 36 23 34 22 32"
```

Mean (or average) values are the basic representational language for models when using k-means. The text view provides a summary of the mean value of each variable over the whole dataset of observations (with the output truncated here):

```
Data means:


        MinTemp          MaxTemp        Rainfall    Evaporation
          7.146           20.372           1.377          4.544
       Sunshine WindGustSpeed  WindSpeed9am   WindSpeed3pm
          8.083           39.944           9.819         18.056
     Humidity9am      Humidity3pm   Pressure9am    Pressure3pm
         71.472           43.859        1019.748       1016.979
```

Cluster Means

A model from a k-means clustering point of view consists of ten (because ten clusters is the default) vectors of the mean values for each of the variables. The main content of the text view is a list of these means. We only show the first five variables and only eight of the ten clusters:

```
Cluster centers:


     MinTemp MaxTemp Rainfall Evaporation Sunshine
1     8.5000   21.05  1.27826       6.330   10.496
2    11.6059   30.95  0.11765       7.647   11.276
3    13.4136   28.77  1.02727       6.200    9.464
4     9.1818   16.90  4.94545       3.800    2.191
5     7.7412   15.19  3.58824       3.306    5.659
6     2.7667   17.16  0.66111       2.656    7.689
7    -0.7913   13.71  0.03478       1.922    7.496
8    11.3088   26.37  0.50000       6.288   10.259
9     1.5045   17.55  0.23636       3.500   10.223
10    7.8625   17.60  2.21875       4.519    6.122
```

Model Quality

The means are followed by a simple measure of the quality of the model:

```
Within cluster sum of squares:

 [1] 14460   5469   8062 11734 11062   9583   7258   7806   6146
[10] 11529
```

The measure used is the sum of the squares of the differences between the observations within each of the ten clusters.

Time Taken

Finally, we see how long the k-means algorithm took to build the ten clusters. For such a small dataset, very little time is required. The time taken is the amount of CPU time spent on the task:

```
Time taken: 0.00 secs
```

Tuning Options

The basic tuning option for building a k-means model in **Rattle** is simply the **Number of clusters** that are to be generated. The default is 10, but any positive integer greater than 1 is allowed.

Rattle also provides an option to iteratively build more clusters and measure the quality of each resulting model as a guide to how many clusters to build. This is chosen by enabling the **Iterate Clusters** option. When active, a model with two clusters, then a model with three clusters, and so on up to a model with ten (or as many as specified) clusters will be built. A plot is generated and displayed to report the improvement in the quality measure (the sum of the within cluster sum of squares).

As mentioned previously, the **Seed** option allows different starting points to be used for the heuristic search. Each time a different seed is used, the resulting model will usually be different.

For some datasets, differences between the models using different seeds will often not be too large, though for others they might be quite large. In the latter case, we are finding different, possibly less optimal or perhaps equally optimal models each time. The **Runs** option will repeat the model building the specified number of times and choose the model that provides the best performance against the measure of model quality. For each different seed, we can check the list of cluster size to confirm that we obtain a collection of clusters that are about the same sizes each time, though the order in the listing changes.

Once a model has been built, the Stats, Data Plot, and Discriminant Plot buttons become available. Clicking the Stats button will result in quite a few additional cluster statistics being displayed in the text view. These can all participate in determining the quality of the model and comparing one k-means model against another. The Data Plot and the Discriminant Plot buttons result in plots that display how the clusters are distributed across the data. The discriminant coordinates plot is generated by projecting the original data to display the key differences between clusters, similar to principal components analysis. The plots are probably only useful for smaller datasets (in the hundreds or thousands).

The Rattle user interface also provides access to the Clara, Hierarchical, and BiCluster clustering algorithms. These are not covered here.

Building a Model Using R

The primary function used within R for k-means clustering is `kmeans()` which comes standard with R. We can build a k-means cluster model using the encapsulation idea presented in Section 2.9:

```
> weatherDS <- new.env()
```

From the *weather* dataset, we will select only two numeric variables on which to cluster, and we also ignore the output variable `RISK_MM`:

```
> library(rattle)
> evalq({
    data <- weather
    nobs <- nrow(data)
  }, weatherDS)
```

We now create a model container to store the results of the modelling and build the actual model. The container also includes the weatherDS dataset information.

```
> weatherKMEANS <- new.env(parent=weatherDS)
> evalq({
    model <- kmeans(x=na.omit(data[, vars]), centers=10)
  }, weatherKMEANS)
```

We have used `kmeans()` and passed to it a dataset with any observations having missing values omitted. The function otherwise complains if the data contains missing values, as we might expect when using a distance measure. The `centers=` option is used either to specify the number of clusters or to list the starting points for the clustering.

9.5   Discussion

Number of Clusters

The primary tuning parameter for the k means algorithm is the number of clusters, k. Simply because the default is to identify ten clusters does not mean that 10 is a good choice at all. Choosing the number of clusters is often quite a tricky exercise. Sometimes it is a matter of experimentation and other times we might have some other knowledge to help us decide. We will soon note that the larger the number of clusters relative to the size of the sample, the smaller our clusters will generally be. However, a common observation is that often we might end up with a small number of clusters containing most of the observations and a large number of clusters containing only a few observations each.

We also note that different cluster algorithms (and even simply using different random seeds to initiate the clustering) can result in different (and sometimes very different) clusters. How much they differ is a measure of the stability of the clustering.

Rattle provides an Iterate Clusters option to assist with identifying a good number of clusters. The approach is to iterate through different values of k. For each k, we observe the sum of the within cluster sum of squares. A plot is generated to show both the sum of squares and its change in the sum of squares. A heuristic is to choose the number of clusters where we see the largest drop in the sum of the within cluster sum of squares.

Shape of Clusters

One of the characteristics to distinguish between clustering algorithms is the shape of the resulting clusters. Essentially, the k-means algorithm, as with any algorithm that uses the distance to a mean as the representation of the clusters, produces convex clusters. Other clustering algorithms exist that can produce differently shaped clusters that might better reflect the data.

Other Cluster Algorithms

R supports a very large variety of clustering algorithms besides the k-means algorithm we have described here. They are grouped into the partitioning type of algorithms, of which k-means is one example, model-based algorithms (see **mclust** (Fraley and Raftery, 2006), for example), and hierarchical clustering (see hclust() from **stats** and agnes() for agglomerative clustering, and diana() for divisive clustering from **cluster** (Maechler et al., 2005)). Rattle supports the building of hierarchical clusters using hclust(). Such an algorithm builds the clusters iteratively and hierarchically.

For an agglomerative hierarchical approach, the two closest observations form the first cluster. Then the next two closest observations, but now also including the mean of the first cluster as a "combined" observation, form the second cluster, and so on until we have formed a single cluster. The resulting collection of potential clusters can be drawn using a dendrogram, as shown in Figure 9.4.

An advantage of this approach is that we get a visual clue as to the number of clusters that naturally appear in the data. In Figure 9.4 we have drawn boxes to indicate perhaps three clusters. A disadvantage is that this approach is really only useful for a small dataset.

Recent research has explored the issue of very high dimensional data, or data with very many variables. For such data the k-means algorithm performs rather poorly, as all observations essentially become equidistant from each other. A successful approach has been developed (Jing et al., 2007) using a weighted distance measure. The algorithm essentially chooses only subsets of the variables on which to cluster. This has been referred to as subspace clustering.

The **siatlust** (Williams et al., 2011) package, provides an implementation of this modification to the k-means algorithm. Entropy weighted variable selection through ewkm() is used to improve the clustering performance with high dimensional data.

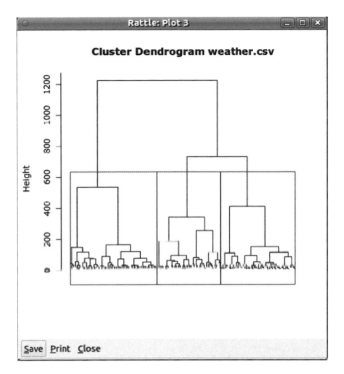

Figure 9.4: A sample dendrogram showing three clusters.

9.6   Command Summary

This chapter has referenced the following R packages, commands, functions, and datasets:

agnes()	function	An agglomerative clustering algorithm.
cluster	package	A variety of tools for cluster analysis.
diana()	function	A divisive clustering algorithm.
ewkm()	function	Entropy weighted k-means.
evalq()	function	Access environment for storing data.
grid()	command	Add a grid to a plot.
hclust()	function	A hierarchical clustering algorithm.
kmeans()	function	The k-means clustering algorithm.
mean	function	Calculate the mean values.

`plot()`	command	Draw a dendrogram for an hclust object.
`round()`	function	Round numbers to specific digits.
`set.seed()`	command	Reset random sequence for sampling.
siatclust	package	Weighted and subspace k-means.
stats	package	Base package providing k-means.
weather	dataset	Sample dataset from **rattle**.

Chapter 10

Association Analysis

Many years ago, a number of new Internet businesses were created to sell books on-line. Over time, they collected information about the books that each of their customers were buying. Using association analysis, they were able to identify groups of books that customers with similar interests seem to have been buying. Using this information, they were able to develop recommendation systems that informed their customers that other customers who purchased some book of interest also purchased other related books.

 *and*

The customer would often find such recommendations quite useful.

Association analysis identifies relationships or correlations between observations and/or between variables in our datasets. These relationships are then expressed as a collection of so-called association rules. The approach has been particularly successful in mining very large transactional databases, like shopping baskets and on-line customer purchases. Association analysis is one of the core techniques of data mining.

For the on-line bookselling example, historic data is used to identify, for example, that customers who purchased two particular books also tended to purchase another particular book. The historic data might indicate that the first two books are purchased by only 0.5% of all customers. But 70% of these then also purchase the third book. This is an *interesting* group of customers. As a business, we will take advantage of this observation by targeting advertising of the third book to those customers who have purchased both of the other books.

The usual type of input data for association analysis consists of transactional data, such as the items in a shopping basket at a supermarket, books and videos purchased by a single client, or medical treatments and tests received by patients. We are interested in the items whose co-occurrence within a transaction is of interest.

This short chapter introduces the basic concepts of association analysis and how to perform it in both **Rattle** and R.

10.1   Knowledge Representation

A representation of association rules is required to identify relationships between items within transactions. Suppose each transaction is thought of as a basket of items (which we might represent as $\{A, B, C, D, E, F\}$). The aim is to identify collections of items that appear together in multiple baskets (e.g., perhaps the items $\{A, C, F\}$ appear together in quite a few shopping baskets). From these so called *itemsets* (i.e., sets of items) we identify rules like $A, F \Rightarrow C$ that tell us that when A and F appear in a transaction (e.g., a shopping basket) then typically so does C.

A collection of association rules then represents a model as the outcome of association analysis. The general format of an association rule is

$$\mathcal{A} \rightarrow \mathcal{C}.$$

Both $\mathcal{A}$ (the left hand side or antecedent) and $\mathcal{C}$ (the right side or consequent) are sets of items. Generally we think of items as being particular books, for example, or particular grocery items. Examples might be:

$$milk \rightarrow bread,$$

$$beer \; \& \; nuts \rightarrow potato \; crisps,$$

$$Shrek1 \rightarrow Shrek2 \; \& \; Shrek3.$$

The concept of an item can be generalised to a specific variable/value combination as the item. The concept of association analysis can then be applied to many different datasets. Using our *weather* dataset, for example, this representation will lead to association rules like

$$WindDir3pm = NNW \rightarrow RainToday = No.$$

10.2   Search Heuristic

The basis of an association analysis algorithm is the generation of frequent itemsets. A frequent itemset is a set of items that occur together frequently enough to be considered as a candidate for generating association rules.

The obvious approaches to identifying itemsets that appear frequently enough in the data are quite expensive computationally, even with moderately sized datasets. The apriori algorithm takes advantage of the simple observation that all subsets of a frequent itemset must also be frequent. That is, if $\{milk, bread, cheese\}$ is a frequent itemset then so must each of the smaller itemsets, $\{milk, bread\}$, $\{milk, cheese\}$, $\{bread, cheese\}$, $\{milk\}$, $\{bread\}$, and $\{cheese\}$. This observation allows the algorithm to consider a significantly reduced search space by starting with frequent individual items. This first step eliminates very rare items. We then combine the remaining single items into itemsets containing just two items and retain only those that are frequent enough and similarly for itemsets containing three items and so on.

The concept of *frequent enough* is a parameter of the algorithm used to control the number of association rules discovered. This is called the *support* and specifies how frequently the items must appear in the whole dataset before they can be considered as a candidate association rule. For example, the user may choose to consider only sets of items that occur in at least 5% of all transactions.

The second phase of the algorithm considers each of the frequent itemsets and for each generates all possible combinations of association rules. Thus, for an itemset containing three items $\{milk, bread, cheese\}$, the following are among the possible association rules that will be considered:

$$bread \ \& \ milk \rightarrow cheese,$$

$$milk \rightarrow bread \ \& \ cheese,$$

$$cheese \ \& \ milk \rightarrow bread,$$

and so on.

The actual association rules that we retain are those that meet a criterion called *confidence*. The confidence calculates the proportion of transactions containing $\mathcal{A}$ that also contain $\mathcal{C}$. The confidence specifies a minimal probability for the association rule. For example, the user may choose to generate only rules that are true at least 90% of the time (that is, when $\mathcal{A}$ appears in the basket, $\mathcal{C}$ also appears in the same basket at least 90% of the time).

The apriori algorithm is a breadth-first or generate-and-test type of search algorithm. Only after exploring all of the possibilities of associations containing k items does it then consider those containing $k + 1$ items. For each k, all candidates are tested to determine whether they have enough support.

In summary, the algorithm uses a simple two-phase generate-and-merge process. In phase 1, we generate frequent itemsets of size k (iterating from 1 until we have no frequent k-itemsets) and then combine them to generate candidate frequent itemsets of size $k + 1$. In phase 2, we build candidate association rules.

10.3   Measures

The two primary measures used in association analysis are the **support** and the **confidence**. The minimum support is expressed as a percentage of the total number of transactions in the dataset. Informally, it is simply how often the items appear together from amongst all of the transactions. Formally, we define *support* for a collection of items $\mathcal{I}$ as the proportion of all transactions in which all items in $\mathcal{I}$ appear and express the support for an association rule as

$$support(\mathcal{A} \rightarrow \mathcal{C}) = P(\mathcal{A} \cup \mathcal{C}).$$

Typically, we use small values for the support, since overall the items that appear together "frequently" enough that are of interest generally won't be the obvious ones that regularly appear together.

The minimum confidence is also expressed as the proportion of the total number of transactions in the dataset. Informally, it is a measure of how often the items $\mathcal{C}$ appear whenever the items $\mathcal{A}$ appear in a transaction. Formally, it is a conditional probability:

$$confidence(\mathcal{A} \rightarrow \mathcal{C}) = P(\mathcal{C}|\mathcal{A}) = P(\mathcal{A} \cup \mathcal{C})/P(\mathcal{A}).$$

It can also be expressed in terms of the *support*:

$$confidence(\mathcal{A} \rightarrow \mathcal{C}) = support(\mathcal{A} \rightarrow \mathcal{C})/support(\mathcal{A}).$$

Typically, this measure will have larger values since we are looking for the association rules that are quite strong, so that if we find the items in $\mathcal{A}$ in a transaction then there is quite a good chance of also finding $\mathcal{C}$ in the transaction.

There are a collection of other measures that are used with association rule analysis. One that is used in R and hence Rattle is the **lift**. The lift is the increased likelihood of $\mathcal{C}$ being in a transaction if $\mathcal{A}$ is included in the transaction. It is calculated as

$$lift(\mathcal{A} \rightarrow \mathcal{C}) = confidence(\mathcal{A} \rightarrow \mathcal{C})/support(\mathcal{C}).$$

Another measure is the **leverage**, which captures the fact that a higher frequency of $\mathcal{A}$ and $\mathcal{C}$ with a lower lift may be interesting:

$$leverage(\mathcal{A} \rightarrow \mathcal{C}) = support(\mathcal{A} \rightarrow \mathcal{C}) - support(\mathcal{A}) * support(\mathcal{C}).$$

10.4 Tutorial Example

Two types of association rules were identified above, corresponding to the type of data made available. The simplest case, known as market basket analysis, is when we have a transaction dataset that records just a transaction identifier. The identifier might identify a single shopping basket containing multiple items from shopping or a particular customer or patient and their associated purchases or medical treatments over time. A simple example of a market basket dataset might record the purchases of DVDs by customers (three customers in this case):

```
ID,Item
1,Sixth Sense
1,LOTR1
1,Harry Potter1
1,Green Mile
1,LOTR2
2,Gladiator
2,Patriot
2,Braveheart
3,LOTR1
3,LOTR2
```

The resulting model will then be a collection of association rules that might include

$$LOTR1 \rightarrow LOTR2.$$

The second form of association rule uses a dataset that we are more familiar with. This approach treats each observation as a transaction and the variables as the items in the "shopping basket." Considering the *weather* dataset, we might obtain models that include rules of the form

$$Humidity3pm = High \ \& \ Pressure3pm = Low \rightarrow RainToday = Yes.$$

Both forms are supported in Rattle and R.

Building a Model Using Rattle

Rattle builds association rule models through the Associate tab. The two types of association rules are supported and the appropriate type is chosen using the Baskets check button. If the button is checked then Rattle will use the Ident and Target variables for the analysis, performing a market basket analysis. If the button is not checked, then Rattle will use the Input variables for a rules analysis.

For a basket analysis, the data is thought of as representing shopping baskets (or any other type of collection of items, such as a basket of medical tests, a basket of medicines prescribed to a patient, a basket of stocks held by an investor, and so on). Each basket has a unique identifier, and the variable specified as an Ident variable on the Data tab is taken as the identifier of a shopping basket. The contents of the basket are then the items contained in the column of data identified as the Target variable. For market basket analysis, these are the only two variables used.

To illustrate market basket analysis with Rattle, we can use a very simple and trivial dataset consisting of the DVD movies purchased by customers. The data is available as a CSV file (named dvdtrans.csv) from the Rattle package. The simplest way to load this dataset into Rattle is to first load the default sample *weather* dataset from the weather.csv file into Rattle. We do this by clicking the Execute button on starting Rattle. Then click the Filename button (which will now be showing weather.csv) to list the contents of Rattle's sample CSV folder. Choose dvdtrans.csv and click Open and then Execute. The ID variable will

automatically be chosen as the Ident, but we will need to change the role
of Item to be Target, as in Figure 10.1.

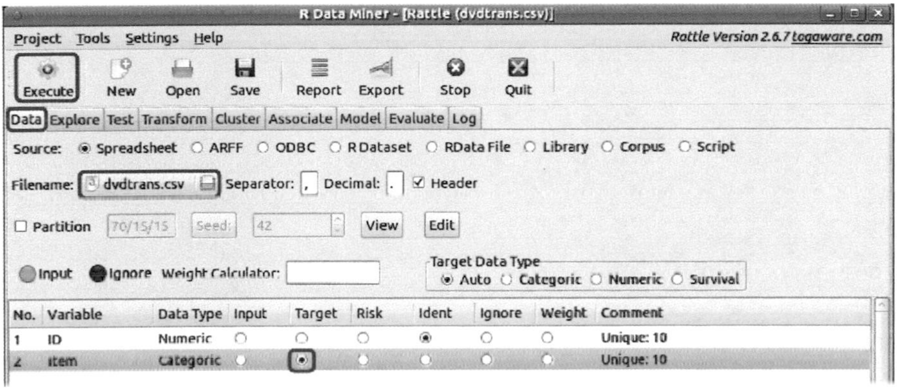

Figure 10.1: Choose the dvdtrans.csv file and load it into Rattle with a
click of the Execute button. Then set the role for Item to be Target and click
Execute for the new role to be noted.

On the Associate tab, ensure that the Baskets button is checked. Click
the Execute button to build a model that will consist of a collection of
association rules. Figure 10.2 shows the resulting text view, which we
now review.

The first few lines of the text view list the number of association rules
that make up the model. In our example, there are 127 rules:

```
Summary of the Apriori Association Rules.


Number of Rules: 127
```

The next code block reports on the distribution of the three measures as
found for the 127 rules of the model:

```
Summary of the Measures of Interestingness:


    support           confidence            lift
Min.    :0.100   Min.    :0.100   Min.    : 0.714
1st Qu.:0.100   1st Qu.:0.500   1st Qu.: 1.429
Median :0.100   Median :1.000   Median : 2.500
Mean    :0.145   Mean    :0.759   Mean    : 3.015
3rd Qu.:0.100   3rd Qu.:1.000   3rd Qu.: 5.000
Max.    :0.700   Max.    :1.000   Max.    :10.000
```

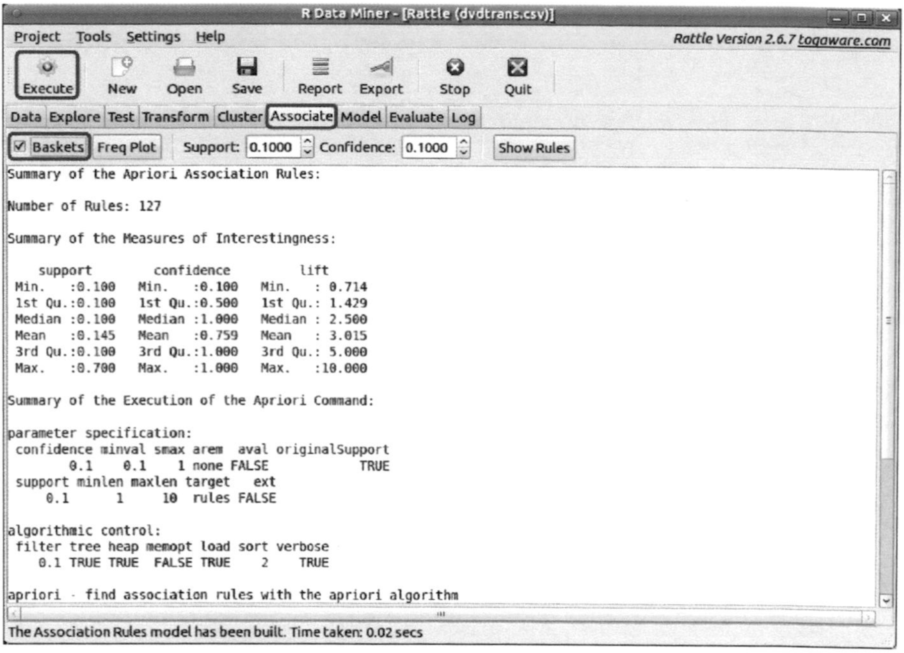

Figure 10.2: Building an association rules model.

The 127 association rules met the criteria of having a minimum support of 0.1 and a minimum confidence of 0.1. Across the rules the support ranges from 0.1 up to 0.4. Confidence ranges from 0.1 up to 1.0 and lift from 0.83 up to 10.0.

This section is followed by a summary of the process of building the model. It begins with a review of the options supplied or the default values for the various parameters. We can see confidence= and support= listed:

```
Summary of the Execution of the Apriori Command:

parameter specification:
 confidence minval smax arem  aval originalSupport
       0.1    0.1    1 none FALSE            TRUE
 support minlen maxlen target   ext
     0.1      1     10  rules FALSE
```

These options are tunable through the Rattle interface. Others can be tuned directly through R. A set of parameters that control how the algorithm itself operates is then displayed:

```
algorithmic control:
 filter tree heap memopt load sort verbose
   0.1 TRUE TRUE  FALSE TRUE    2    TRUE
```

The final section includes detailed information about the algorithm and the model that has been built:

```
apriori - find association rules with the apriori algorithm
version 4.21 (2004.05.09) (c) 1996-2004   Christian Borgelt
set item appearances ...[0 item(s)] done [0.00s].
set transactions ...[10 item(s). 10 trans] done [0.00s].
sorting and recoding items ... [10 item(s)] done [0.00s].
creating transaction tree ... done [0.00s].
checking subsets of size 1 2 3 4 5 done [0.00s].
writing ... [127 rule(s)] done [0.00s].
creating S4 object  ... done [0.00s].
```

The Show Rules button will show all of the association rules for the model in the text view window, sorted by the level of confidence in the rule. The top five rules will be:

	lhs		rhs	supp	conf	lift
1	{Harry Potter2}	=>	{Harry Potter1}	0.1	1	5.000
2	{Braveheart}	=>	{Patriot}	0.1	1	1.667
3	{Braveheart}	=>	{Gladiator}	0.1	1	1.429
4	{LOTR}	=>	{Green Mile}	0.1	1	5.000
5	{LOTR}	=>	{Sixth Sense}	0.1	1	1.667

These rules have only a single item on each side of the arrow, and all have a support of 0.1 and a confidence of 1. We can see that for either of the first two movies there is quite a large lift obtained.

Building a Model Using R

Arules (Hahsler et al., 2011) provides `apriori()` for R. The package provides an interface to the widely used, and freely available, apriori software from Christian Borgelt. This software was, for example, commercially licensed for use in the Clementine[1] data mining package and is a well-developed and respected implementation.

When loading a dataset to process with `apriori()`, it needs to be converted into a transaction data structure. Consider a dataset with two columns, one being the identifier of the "basket" and the other being an item contained in the basket, as is the case for the `dvdtrans.csv` data. We can load that data into R:

```
> library(arules)
> library(rattle)
> dvdtrans <- read.csv(system.file("csv", "dvdtrans.csv",
                                    package="rattle"))
> dvdDS <- new.env()
> dvdDS$data <- as(split(dvdtrans$Item, dvdtrans$ID),
                 "transactions")
> dvdDS$data

transactions in sparse format with
 10 transactions (rows) and
 10 items (columns)
```

We can then build the model using this transformed dataset:

```
> dvdAPRIORI <- new.env(parent=dvdDS)
> evalq({
    model <- apriori(data, parameter=list(support=0.2,
                                 confidence=0.1))
  }, dvdAPRIORI)
```

The rules can be extracted and ordered by confidence using `inspect()`. In the following code block we also use [1:5] to limit the display to just the first five association rules. We notice that the first two are symmetric, which is expected since everyone who purchases one of these movies always also purchases the other.

---

[1] Clementine became an SPSS product and was then purchased by IBM to become IBM SPSS Modeler.

```
> inspect(sort(dvdAPRIORI$model, by="confidence")[1:5])

  lhs                  rhs              support confidence  lift
1 {LOTR1}          => {LOTR2}              0.2           1 5.000
2 {LOTR2}          => {LOTR1}              0.2           1 5.000
3 {Green Mile}     => {Sixth Sense}        0.2           1 1.667
4 {Patriot}        => {Gladiator}          0.6           1 1.429
5 {Patriot,
    Sixth Sense}  => {Gladiator}          0.4           1 1.429
```

10.5   Command Summary

This chapter has referenced the following R packages, commands, functions, and datasets:

apriori()	function	Build an association rule model.
arules	package	Support for association rules.
inspect()	function	Display results of model building.
weather	dataset	Sample dataset from **rattle**.

Chapter 11

Decision Trees

Decision trees (also referred to as classification and regression trees) are the traditional building blocks of data mining and the classic machine learning algorithm. Since their development in the 1980s, decision trees have been the most widely deployed machine-learning based data mining model builder.

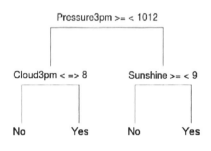

Their attraction lies in the simplicity of the resulting model, where a decision tree (at least one that is not too large) is quite easy to view, understand, and, importantly, explain. Decision trees do not always deliver the best performance, and represent a trade-off between performance and simplicity of explanation. The decision tree structure can represent both classification and regression models.

We introduce the decision tree as a knowledge representation language in Section 11.1. A search algorithm for finding a good decision tree is presented in Section 11.2. The measures used to identify a good tree are discussed in Section 11.3. Section 11.4 then illustrates the building of a decision tree in Rattle and directly through R. The options for building a decision tree are covered in Section 11.5.

11.1   Knowledge Representation

The tree structure is used in many different fields, such as medicine, logic, problem solving, and management science. It is also a traditional computer science structure for organising data. We generally present the tree upside down, with the *root* at the top and the leaves at the bottom. Starting from the root, the tree splits from the single trunk into two or more branches. Each branch itself might further split into two or more branches. This continues until we reach a leaf, which is a node that is not further split. We refer to the split of a branch as a *node* of the tree. The root and leaves are also referred to as *nodes*.

A *decision tree* uses this traditional structure. It starts with a single root node that splits into multiple branches, leading to further nodes, each of which may further split or else terminate as a leaf node. Associated with each nonleaf node will be a test or question that determines which branch to follow. The leaf nodes contain the "decisions."

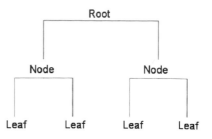

Consider the decision tree drawn on page 205 (which is the same tree as Figure 2.5 on page 30). This represents knowledge about observing weather conditions one day and the observation of rain on the following day. The No and Yes values at the leaves of the decision tree represent the decisions.

The root node of the example decision tree tests the mean sea level pressure at 3 pm (Pressure3pm). When this variable, for an observation, has a value greater than or equal to 1012 hPa, then we will continue down the left side of the tree. The next test down this left side of the tree is on the amount of cloud cover observed at 3 pm (Cloud3pm). If this is less than 8 oktas (i.e., anything but a fully overcast sky), then it is observed that on the following day it generally does not rain (No). If we observe that it is overcast today at 3 pm (i.e., Cloud3pm is 8 oktas, the maximum value of this variable—see Section 5.2.9, page 127) then generally we observe that it rains the following day (Yes). Thus we would be inclined to think that it might rain tomorrow if we observe these same conditions today.

Resuming our interpretation of the model from the root node of the tree, if `Pressure3pm` is less than 1012 hPa and `Sunshine` is greater than or equal to 9 (i.e., we observe at least 9 hours of sunshine during the day), then we do not expect to observe rain tomorrow. If we record 9 or less hours of sunshine, then we expect it to rain tomorrow.

The decision tree is a very convenient and efficient representation of knowledge. Generally, models expressed in one language can be translated to another language—and so it is with a decision tree. One simple and useful translation is into a rule set. The decision tree above translates to the following rules, where each rule corresponds to one pathway through the decision tree, starting at the root node and terminating at a leaf node:

```
Rule number: 7 [RainTomorrow=Yes cover=27 (11%) prob=0.74]
   Pressure3pm< 1012
   Sunshine< 8.85

Rule number: 5 [RainTomorrow=Yes cover=9 (4%) prob=0.67]
   Pressure3pm>=1012
   Cloud3pm>=7.5

Rule number: 6 [RainTomorrow=No cover=25 (10%) prob=0.20]
   Pressure3pm< 1012
   Sunshine>=8.85

Rule number: 4 [RainTomorrow=No cover=195 (76%) prob=0.05]
   Pressure3pm>=1012
   Cloud3pm< 7.5
```

A rule representation has its advantages. In reviewing the knowledge that has been captured, we can consider each rule separately rather than being distracted by the more complex structure of a large decision tree. It is also easy to see how each rule could be translated into a programming language statement like R, Python, C, VisualBasic, or SQL. The structure is as simple, and clear, as an If-Then statement. We now explain the information provided for each rule.

In building a decision tree, often a larger tree is built and then cut back (or pruned) so that it is not so complex and also to improve its accuracy. As a consequence, we will often see node numbers (and rule numbers) that are not sequential. The node numbers do not have any specific meaning other than as a reference.

Although it is not shown in the tree representation at the beginning of the chapter, we see in the rules above the probabilities that are typically recorded for each leaf node of the decision tree. The probabilities can be used to provide an indication of the strength of the decision we derive from the model. Thus, rule number 7 indicates that for 74% of the observations (`prob=0.74`), when the observed pressure at 3 pm is less than 1012 hPa and the hours of sunshine are less than 8.85 hours, there is rainfall recorded on the following day (`RainTomorrow=Yes`). The other information provided with the rule is that 27 observations from the training dataset (i.e., 11% of the training dataset observations) are covered by this rule—they satisfy the two conditions.

There exist variations to the basic decision tree structure we have presented here for representing knowledge. Some approaches, as here, limit trees to two splits at any one node to generate a *binary decision tree*. For categoric data this might involve partitioning the values (levels) of the variable into two groups. Another approach is to have a branch corresponding to each of the levels of a categoric variable. From a representation point of view, what can be represented using a multiway tree can also be represented as a binary tree and vice versa. Other variations, for example, allow multiple variables to be tested at a node. We generally stay with the simpler representation, though, sometimes at the cost of the resulting model being a little more complex than if we used a more complex decision tree structure.

11.2   Algorithm

Identifying Alternative Models

The decision tree structure, as described above, is the "language" we use to express our knowledge. A sentence (or model) in this language is a particular decision tree. For any dataset, there will be very many, or even infinite, possible decision trees (sentences).

Consider the simple decision tree discussed above. Instead of the variable `Pressure3pm` being tested against the value 1012, it could have been tested against the value 1011, or 1013, or 1020, etc. Each would, when the rest of the tree has been built, represent a different sentence in the language, representing a slightly different capture of the knowledge. There are very many possible values to choose from for just this one variable, even before we begin to consider values for the other variables that appear in the decision tree.

Alternatively, we might choose to test the value of a different variable at the root node (or any other node). Perhaps we could test the value of Humidity3pm instead of Pressure3pm. This again introduces a large collection of alternative sentences that we might generate within the constraints of the language we have defined. Each sentence is a candidate for the capture of knowledge that is consistent with the observations represented in our training dataset.

As we saw in Section 8.2, this wealth of possible sentences presents a challenge—which is the best sentence or equivalently which is the best model that fits the data? Our task is to identify the sentence (or perhaps sentences) that best captures the knowledge that can be obtained from the observations that we have available to us.

We generally have an infinite collection of possible sentences to choose from. Enumerating every possible sentence, and testing whether it is a good model, will generally be too computationally expensive. This could well involve days, weeks, months, or even more of our computer time. Our task is to use the observations (the training dataset) to narrow down this search task so that we can find a good model in a reasonable amount of time.

Partitioning the Dataset

The algorithm that has been developed for decision tree induction is referred to as the top-down induction of decision trees, using a divide-and-conquer, or recursive partitioning, approach. We will describe the algorithm intuitively.

We continue here with the *weather* dataset to describe the algorithm. The distribution of the observations, with respect to the target variable RainTomorrow, is of particular interest. There are 66 observations that have the target as Yes (18%) and 300 observations with No (82%).

We want to find any input variable that can be used to split the dataset into two smaller datasets. The goal is to increase the homogeneity of each of the two datasets with respect to the target variable. That is, for one of the datasets, we would be looking for it to have an increased proportion of observations with Yes and so the other dataset would have an increased proportion of observations with No.

We might, for example, decide to construct a partition of the
original dataset using the variable
Sunshine with a split value of 9.
Every observation that has a value
of Sunshine less than 9 goes into

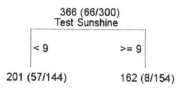

one subset and those remaining (with Sunshine equal to 9) into a second subset. These new datasets will have 201 and 162 observations,
respectively (noting that three observations have missing values for this
variable).

Now we consider the proportions of Yes and No observations within
the two new datasets. For the subset of observations with Sunshine less
than 9, the proportions are 28% Yes and 72% No. For the subset of
observations with Sunshine greater than or equal to 9 the proportions
are 5% Yes and 95% No.

By splitting on this variable, we have made an improvement in the
homogeneity of the target variable values. In particular, the right dataset
(Sunshine ≥ 9) results in a collection of observations that are very much
in favour of no rain on the following day (95% No). This is what we are
aiming to do. It allows us to observe that when the amount of sunshine
on any day is quite high (i.e., at least 9 hours), then there is very little
chance of rain on the following day (only a 5% chance based on our
observations from the particular weather station).

The story for the other dataset is not quite so clear. The proportions
have certainly changed, with a higher proportion of Yes observations
than the original dataset, but the No observations still outnumber the Yes
observations. Nonetheless, we can say that when we observe Sunshine <
9 there is an increased likelihood of rain the following day based on our
historic observations. There is a 28% chance of rain compared with an
18% over all observations.

Choosing the value 9 for the variable Sunshine is just one possibility from amongst very many choices. If we had chosen the value 5 for
the variable Sunshine we would have two new datasets with the Yes/No
proportions 41%/59% and 12%/88%. Choosing a different variable altogether (Cloud3pm) with a split of 6, we would have two new datasets with
the Yes/No proportions 8%/92% and 34%/66%. Another choice might
be Pressure3pm with a split of 1012. This gives the Yes/No proportions
as 47%/53% and 10%/90%.

We now have a collection of choices for how we might partition our training dataset: which of these is the best split? We come back to answer that question formally in Section 11.3. For now, we assume we choose one of them. With whichever choice we make, the result is that we now have two new smaller datasets.

Recursive Partitioning

The process is now repeated again separately for the two new datasets. That is, for the left dataset above (observations having Sunshine < 9), we consider all possible variables and splits to partition that dataset into two smaller datasets. Independently, for the right dataset (observations having Sunshine ≥ 9) we consider all possible variables and splits to partition that dataset into two smaller datasets as well.

Now we have four even smaller datasets—and the process continues. For each of the four datasets, we again consider all possible variables and splits, choosing the "best" at each stage, partitioning the data, and so on, repeating the process until we decide that we 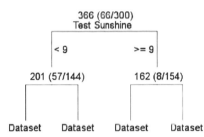 should stop. In general, we might stop when we run out of variables, run out of data, or when partitioning the dataset does not improve the proportions or the outcome.

We can see now why this process is called divide-and-conquer or recursive partitioning. At each step, we have identified a question, that we use to partition the data. The resulting two datasets then correspond to the two branches of the tree emanating from that node. For each branch, we identify a new question and partition appropriately, building our representation of the knowledge we are discovering from the data. We continually divide the dataset and conquer each of the smaller datasets more easily. We are also repeatedly partitioning the dataset and applying the same process, independently, to each of the smaller datasets; thus it is recursive partitioning.

At each stage of the process, we make a decision as to the best variable and split to partition the data. That decision may not be the best to make in the overall context of building this decision tree, but once we make that decision, we stay with it for the rest of the tree. This is generally referred to as a *greedy* approach.

A greedy algorithm is generally quite efficient, whilst possibly sacrificing our opportunity to find the very best decision tree. There remains quite a bit of searching for the one variable and split point for each of the datasets we produce. However, this heuristic approach reduces our search space considerably by fixing the variable/split once it has been chosen.

11.3   Measures

In describing the basic algorithm above, it was indicated that we need to measure how good a particular partition of the dataset is. Such a measure will allow us to choose from amongst a collection of possibilities. We now consider how to measure the different splits of the dataset.

Information Gain

Rattle uses an *information gain* measure for deciding between alternative splits. The concept comes from information theory and uses a formulation of the concept of entropy from physics (i.e., the concept of the amount of disorder in a system). We discuss the concepts here in terms of a binary target variable, but the concept generalises to multiple classes and even to numeric target variables for regression tasks.

For our purposes, the concept of disorder relates to how "mixed" our dataset is with respect to the values of the target variable. If the dataset contains only observations that all have the same value for the target variable (e.g., it contains only observations where it rains the following day), then there is no disorder—i.e., no entropy or zero entropy. If the two values of the target variable are equally distributed across the observations (i.e., 50% of the dataset are observations where it rains tomorrow and the other 50% are observations where it does not rain tomorrow), then the dataset contains the maximum amount of disorder. We identify the maximum amount of entropy as 1. Datasets containing different mixtures of the values of the target variable will have a measure of entropy between 0 and 1.

From an information theory perspective, we interpret a measure of 0 (i.e., an entropy of 0) as indicating that we need no further information in order to classify a specific observation within the dataset—all observations belong to the same class. Conversely, a measure of 1 suggests we need the maximal amount of extra information in order to classify our observations into one of the two available classes. If the split between the observations where it rains tomorrow and where it does not rain tomorrow is not 50%/50% but perhaps 75%/25%, then we need less extra information in order to classify our observations—the dataset already contains some information about which way the classification is going to go. Like entropy, our measure of "required information" is thus between 0 and 1.

In both cases, we will use the mathematical logarithm function for base 2 ($\log_2$) to transform our proportions (the proportions being 0.5, 0.75, 1.00, etc.). Base 2 is chosen since we use binary digits (bits) to encode information. However, we can use any base since in the end it is the relative measure rather than the exact measure, that we are interested in and the logarithm functions have identical behaviour in this respect. The default R implementation (as we will see in Section 11 4) uses the natural logarithm, for example.

The formula we use to capture the entropy of a dataset, or equivalently the information needed to classify an observation, is

$$info(\mathcal{D}) = -p\log_2(p) - n\log_2(n)$$

We now delve into the nature of this formula to understand why this is a useful measure. We can easily plot this function, as in Figure 11.1, with the x-axis showing the possible values of p and the y-axis showing the values of $info$.

From the plot, we can see that the maximum value of the measure is 1. This occurs when there is the most amount of disorder in the data or when the most amount of additional information is required to classify an observation. This occurs when the observations are equally distributed across the values of the target variable. For a binary target, as here, this occurs when $p = 0.5$ and $n = 0.5$.

Likewise, the minimum value of the measure is 0. This occurs at the extremes, where $p = 1$ (i.e., all observations are positive —RainTomorrow has the value Yes for each) or $p = 0$ (i.e., all observations are negative —RainTomorrow has the value No for each). This is interpreted as either no entropy or as requiring no further information in order to classify the observations.

This then provides a mechanism for measuring some aspect of the training dataset, capturing something about the knowledge content. As

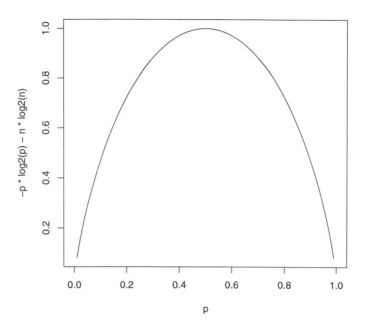

Figure 11.1: Plotting the relationship between the proportion of positive observations in the data and the measure of information/entropy.

we now see, we use this formulation to help choose the "best" split from among the very many possible splits we identified in Section 11.2.

Each choice of a split results in a binary partition of the training dataset. We will call these $\mathcal{D}_1$ and $\mathcal{D}_2$, noting that $\mathcal{D} = \mathcal{D}_1 \cup \mathcal{D}_2$. The information measure can be applied to each of these subsets to give $\mathcal{I}_1$ and $\mathcal{I}_2$. If we add these together, weighted by the sizes of the two subsets, we get a measure of the combined information, or entropy:

$$info(\mathcal{D}, \mathcal{S}) = \frac{|\mathcal{D}_1|}{|\mathcal{D}|}\mathcal{I}_1 + \frac{|\mathcal{D}_2|}{|\mathcal{D}|}\mathcal{I}_2$$

Comparing this with the original information, or entropy, we get a measure of the gain in "knowledge" obtained by using the particular split point:

$$gain(\mathcal{D}, \mathcal{S}) = info(\mathcal{D}) - info(\mathcal{D}, \mathcal{S})$$

This can then be calculated for each of the possible splits. The split that provides the greatest gain in information (and equivalently the greatest reduction in entropy) is the split we choose.

Other Measures

A variety of measures can be used as alternatives to the information measure. The most common alternative is the Gini index of diversity. This was introduced into decision tree building through the original CART (classification and regression tree) algorithm (Breiman et al., 1984). The plot of the function is very similar to the $p * \log_2(p)$ curve and typically will give the same split points.

11.4   Tutorial Example

The *weather* dataset is used to illustrate the building of a decision tree. We saw our first decision tree in Chapter 2. We can build a decision tree using Rattle's Tree option, found on the Model tab or directly in R through rpart() of **rpart** (Therneau and Atkinson, 2011).

Building a Model Using Rattle

We build a decision tree using Rattle's Model tab's Tree option. After loading our dataset and identifying the Input variables and the Target variable, an Execute of the Model tab will result in a decision tree. We can see the result for the *weather* dataset in Figure 11.2, which shows the resulting tree in the text view and also highlights the key interface widgets that we need to deal with to build a tree.

The text view includes much information, and we will work our way through its contents. However, before doing so, we can get a quick view of the resulting decision tree by using the Draw button of the interface. A window will pop up, displaying the tree, as we saw in Figure 2.5 on page 30.

Working our way through the textual summary of the decision tree, we start with a report of the number of observations that were used to build the tree (i.e., 256):

```
Summary of the Decision Tree model for Classification ...

n= 256
```

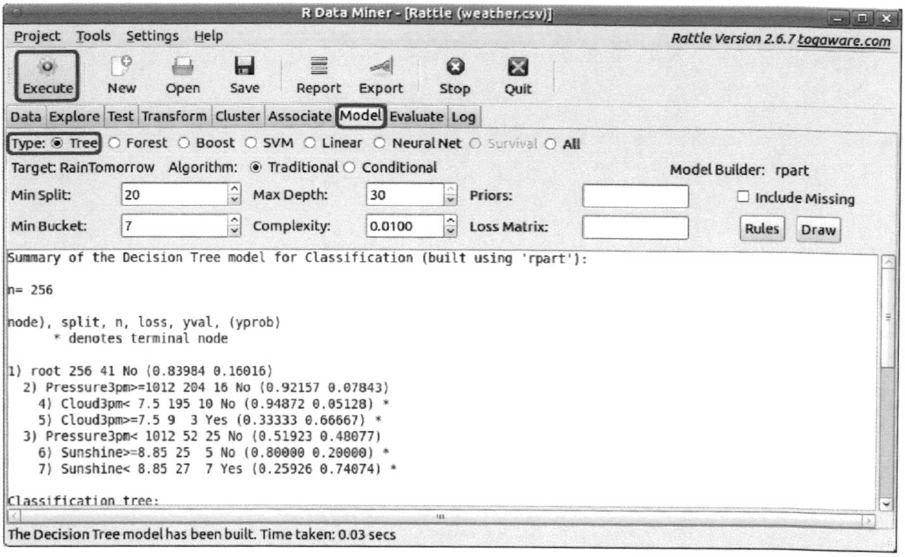

Figure 11.2: Building a decision tree predictive model using the *weather* dataset.

Tree Structure

We now look at the structure of the tree as it is presented in the text view. A legend is provided to assist in reading the tree structure:

```
node), split, n, loss, yval, (yprob)
      * denotes terminal node
```

The legend indicates that a **node** number will be provided, followed by a **split** (which will usually be in the form of a *variable operation value*), the number of entities **n** at that node, the number of entities that are incorrectly classified (the **loss**), the default classification for the node (the **yval**), and then the distribution of classes in that node (the **yprobs**). The distribution is ordered by class and the order is the same for all nodes. The next line indicates that a "*" denotes a terminal node of the tree (i.e., a leaf node—the tree is not split any further at that node).

The first node of any tree is always the root node. We work our way into the tree itself through the root node. The root node is numbered as node number 1:

```
1) root 256 41 No (0.83984 0.16016)
```

The root node represents all observations. By itself the node represents a model that simply classifies every observation into the class that is associated with the majority from the training dataset. The information provided tells us that the majority class for the root node (the `yval`) is No. The 41 tells us how many of the 256 observations will be incorrectly classified as `Yes`. This is technically called the `loss`.

The `yprob` component then reports on the distribution of the classes across the observations. We know the classes to be No, and Yes. Thus, 84% (i.e., 0.83984375 as a proportion) of the observations have the target variable `RainTomorrow` as No, and 16% of the observations have it as Yes.

If the root node itself were treated as a model, it would always decide that it won't rain tomorrow. Based on the training dataset, the model would be 84% correct. That is quite a good level of accuracy, but the model is not particularly useful since we are really interested in whether it is going to rain tomorrow.

The root node is split into two subnodes. The split is based on the variable `Pressure3pm` with a split value of `1011.9`. Node 2 has the *split* expressed as `Pressure3pm>=1011.9`. That is, there are 204 observations with a 3 pm pressure reading of more than 1011.9 hPa:.

```
2) Pressure3pm>=1012 204 16 No (0.92157 0.07843)
```

Only 16 of these 204 observations are misclassified, with the classification associated with this node being No. This represents an accuracy of 92% in predicting that it does not rain tomorrow.

Node 3 contains the remaining 52 observations which have a 3 pm pressure of less than `1011.9`. Whilst the decision is No, it is pretty close to a 50/50 split in this partition:

```
3) Pressure3pm< 1012 52 25 No (0.51923 0.48077)
```

We've skipped ahead a little to jump to node 3, so we now have a look again at node 2 and its split into subnodes. The algorithm has chosen `Cloud3pm` for the next split, with a split value of `7.5`. Node 4 has 195 observations. These are the 195 observations for which the 3 pm pressure is greater than or equal to 1011.9 and the cloud coverage at 3 pm is less than 7.5. Under these circumstances, there is no rain the following day 95% of the time.

```
2) Pressure3pm>=1012 204 16 No (0.92157 0.07843)
  4) Cloud3pm< 7.5 195 10 No (0.94872 0.05128) *
  5) Cloud3pm>=7.5 9  3 Yes (0.33333 0.66667) *
```

Node 5, at last, predicts that it will rain on the following day—at least based on the available historic observations. There are only nine observations here, and the frequency of observing rain on the following day is 67%. Thus we say there is a 67% probability of rain when the pressure at 3 pm is at least 1011.9 and the cloud cover at 3 pm is at least 7.5. Both node 4 and node 5 are marked with an asterisk (*), indicating that they are terminal nodes—they are not further split. The remaining nodes, 6 and 7, split node 3 using the variable Sunshine and a split point of 8.85:

```
3) Pressure3pm< 1012 52 25 No  (0.51923 0.48077)
  6) Sunshine>=8.85 25  5 No  (0.80000 0.20000) *
  7) Sunshine< 8.85 27  7 Yes (0.25926 0.74074) *
```

Node 3 has almost equal numbers of No and Yes observations (52% and 48%, respectively). However, splitting on the number of hours of sunshine has quite nicely partitioned the observations into two groups that are quite a bit more homogeneous with respect to the target variable. Node 6 represents only a 20% chance of rain tomorrow, whilst node 7 represents a 74% chance of rain tomorrow.

That then is the model that has been built. It is a relatively simple decision tree with just seven nodes and four leaf nodes, with a maximum depth of 2 (in fact, each leaf node is at a depth of exactly 2).

Function Call

The next segment lists the underlying R command line that is used to build the decision tree. This was automatically generated based on the information provided through the interface. We could have directly entered this at the prompt in the R Console:

```
Classification tree:
rpart(formula=RainTomorrow ~ .,data=crs$dataset[crs$train,
  c(crs$input,crs$target)],method="class",
  parms=list(split="information"),
  control=rpart.control(usesurrogate=0,maxsurrogate=0))
```

The *formula* notes that we want to build a model to predict the value of the variable RainTomorrow based on the remainder of the variables in the dataset supplied (notated as the "~ ."). The dataset supplied consists of the crs$dataset data frame indexed to include the rows listed in the variable `crs$train`. This is the training dataset. The columns from 3 to 22, and then column 24, are included in the dataset from which the model is built.

Following the specification of the formula and dataset are the tuning parameters for the algorithm. These are explained in detail in Section 11.5, but we briefly summarise them here. The method used is based on classification. The method for choosing the best split uses the information measure. Surrogates (for dealing with missing values) are not used by default in Rattle.

Variables Used

In general, only a subset of the available variables will be used in the resulting decision tree model. The next segment lists those variables that do appear in the tree. Of the 20 input variables, only three are used in the final model.

```
Variables actually used in tree construction:
[1] Cloud3pm    Pressure3pm Sunshine
```

Performance Evaluation

The next segment summarises the process of building the tree, and in particular the iterations and associated change in the accuracy of the model as new levels are added to the tree. The complexity table is discussed in more detail in Section 11.5.

Briefly, though, we are most likely interested in the cross-validated error (refer to Section 15.1 for a discussion of cross-validation), which is the `xerror` column of the table. The error over the whole dataset (i.e., if we were to classify every observation as No) is 0.16, or 16%. Treating this as the baseline error (i.e., 1.00), the table shows the relative reduction in the error (and cross-validation-based error) as we build the tree.

From line 2, we see that after the first split of the dataset, we have reduced the cross-validation based error to 80% of the original amount (i.e., 0.1289, or 13%). Notice that the cross-validation is being reduced more slowly than the error on the training dataset (`error`). This is typical.

The CP value (the complexity parameter) is explained further in Section 11.5, but for now we note that as the tree splits into more nodes, the complexity parameter is reduced. But we also note that the cross-validation error starts to increase as we further split the decision tree. This tells the algorithm to stop partitioning, as the error rate (at least the unbiased estimate of it—refer to Section 15.1) is not improving:

```
Root node error: 41/256 = 0.16

n= 256

      CP nsplit rel error xerror xstd
1 0.159       0     1.00   1.00 0.14
2 0.073       2     0.68   0.80 0.13
3 0.010       3     0.61   0.95 0.14
```

Time Taken

Finally, we see how long it took to build the tree. Decision trees are generally very quick to build.

```
Time taken: 0.03 secs
```

Tuning Options

The Rattle interface provides a choice of Algorithm for building the decision tree. The Traditional option is chosen by default, and that is what we have presented here. The Conditional option uses a more recent conditional inference tree algorithm, which is explained in more detail in Section 11.6. A variety of other tuning options are also provided, and they are discussed in some detail in Section 11.5.

Displaying Trees

The Rules and Draw buttons provide alternative views of the decision tree. Clicking on the Rules button will translate the decision tree into a set of rules and list those rules at the bottom of the text view. We need to scroll down the text view in order to see the rules. The rules in this form can be more easily extracted and used to generate code in other languages. A common example is to generate a query in SQL to extract the corresponding observations from a database.

The Draw button will pop up a separate window to display a more visually appealing representation of the decision tree. We have seen the pictorial representation of a decision tree a number of times now, and they were generated from this button, as was Figure 11.3.

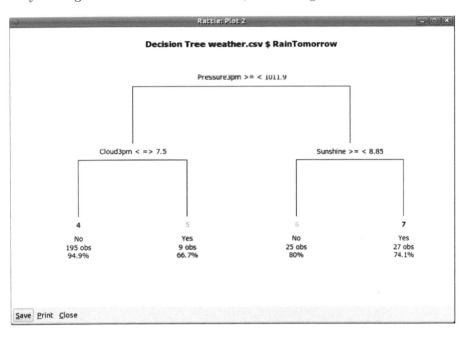

Figure 11.3: Typical Rattle decision tree.

Scoring

We can now use the model to predict the outcome for new observations—something we often call scoring. The Evaluate tab provides the Score option and the choice to Enter some data manually and have that data scored by the model. Executing this setup will result in a popup window in which to enter the data, and, on closing the window, the data is passed on to the model and the predictions are displayed in the Textview.

Building a Model using R

Underneath Rattle's GUI, we are relying on a collection of R commands and functions. The Log tab will expose them, and it is instructive to review the Log tab regularly to gain insight and understanding that will be helpful in using R itself. We effectively lift the bonnet on the hood here so that we can directly build decision trees using R.

To use the traditional decision-tree-building algorithm, we use **rpart**, This provides rpart() which is an implementation of the standard classification and regression tree algorithms. The implementation is very robust and reliable.

```
> library(rpart)
```

As we saw in Section 2.9, we will create the variable weatherDS (using new.env()—new environment) to act as a container for the *weather* dataset and related information. We will access data within this container through the use of evalq() below.

```
> weatherDS <- new.env()
```

The *weather* dataset from **rattle** will be used for the modelling. Three columns from the dataset are ignored in our analyses, as they play no role in the model building. The three variables are the two that serve to identify the observations (Date and Location) and the risk variable (RISK_MM—the amount of rain recorded on the next day). Below we identify the index of these variables and record the negative index in vars, which is stored within the container:

```
> library(rattle)
> evalq({
    data <- weather
    nobs <- nrow(data)
    vars <- -grep('^(Date|Locat|RISK)', names(weather))
  }, weatherDS)
```

A random subset of 70% of the observations is chosen and will be used to identify a training dataset. The random number seed is set, using set.seed(), so that we will always obtain the same random sample for illustrative purposes and repeatability. Choosing different random sample seeds is also useful, providing empirically an indication of how stable the models are.

```
> evalq({
    set.seed(42)
    train <- sample(nobs, 0.7*nobs)
  }, weatherDS)
```

We add to the `weatherDS` container the formula to describe the model that is to be built based on this dataset:

```
> evalq({
    form <- formula(RainTomorrow ~ .)
  }, weatherDS)
```

We now create a model container for the information relevant to the decision tree model that we will build. The container includes the `weatherDS` container (identifying it as `parent=` in the call to `new.env()`):

```
> weatherRPART <- new.env(parent=weatherDS)
```

The command to build a model is then straight forward. The variables `data`, `train`, and `vars` are obtained from the `weatherDS` container, and the result will be stored as the variable `model` within the `weatherRPART` container. We explain `rpart()` in detail below.

```
> evalq({
    model <- rpart(formula=form, data=data[train, vars])
  }, weatherRPART)
```

Here we use `rpart()`, passing to it a formula and the data. We don't need to include the `formula=` and the `data=` in the formal arguments to the function, as they will also be determined from their position in the argument list. It doesn't hurt to include them either to provide more clarity for others reading the code.

The `formula=` argument identifies the model that is to be built. In this case, we pass to the function the variable `form` that we previously defined. The target variable (to the left of the tilde in `form`) is `RainTomorrow`, and the input variables consist of all of the remaining variables in the dataset (denoted by the period to the right of the tilde in `form`). We are requesting a model that predicts a value for `RainTomorrow` based on today's observations.

The `data=` argument identifies the training dataset. Once again, we pass to the function the variable `data` that we previously defined. The training dataset subset consists of the observation numbers listed in the variable `train`. The variables of the dataset that we wish to include are specified by `vars`, which in this case actually lists as negative integers the variables to ignore. Together, `train` and `vars` identify the observations and variables to include in the training of the model.

The result of building the model is assigned into the variable `model` inside the environment `weatherRPART` and so can be independently referred to as `weatherRPART$model`.

Exploring the Model

Towards the end of Section 11.4, we explained the textual presentation of the results of building a decision tree model. The output we saw there can be reproduced in R using `print()` and `printcp()`. The output from `print()` is:

```
> print(weatherRPART$model)

n= 256

node), split, n, loss, yval, (yprob)
      * denotes terminal node

1) root 256 41 No (0.83984 0.16016)
  2) Pressure3pm>=1012 204 16 No (0.92157 0.07843)
    4) Cloud3pm< 7.5 195 10 No (0.94872 0.05128) *
    5) Cloud3pm>=7.5 9  3 Yes (0.33333 0.66667) *
  3) Pressure3pm< 1012 52 25 No (0.51923 0.48077)
    6) Sunshine>=8.85 25  5 No (0.80000 0.20000) *
    7) Sunshine< 8.85 27  7 Yes (0.25926 0.74074) *
```

We briefly discussed the output of this and the `printcp()` below in the previous section. We mentioned there how the CP (complexity parameter) is used to guide how large a decision tree to build. We might choose to stop when the cross-validated error (`xerror`) begins to increase. This is displayed in the output of `printcp()`. We can also obtain a useful graphical representation of the complexity parameter using `plotcp()` instead.

```
> printcp(weatherRPART$model)

Classification tree:
rpart(formula = form, data = data[train, vars])

Variables actually used in tree construction:
[1] Cloud3pm    Pressure3pm Sunshine

Root node error: 41/256 = 0.16

n= 256

      CP nsplit rel error xerror xstd
1 0.159      0      1.00    1.00 0.14
2 0.073      2      0.68    0.83 0.13
3 0.010      3      0.61    0.80 0.13
```

Another command useful for providing information about the resulting model is summary():

```
> summary(weatherRPART$model)
```

This command provides quite a bit more information about the model building process, beginning with the function call and data size. This is followed by the same complexity table we saw above:

```
Call:
rpart(formula=form, data=data[train, vars])
  n= 256

        CP nsplit rel error xerror    xstd
1 0.15854      0    1.0000 1.0000 0.1431
2 0.07317      2    0.6829 0.8293 0.1324
3 0.01000      3    0.6098 0.8049 0.1308
```

The summary goes on to provide information related to each node of the decision. Node number 1 is the root node of the decision tree. Its information appears first (note that the text here is modified to fit the page):

```
Node number 1: 256 observations, complexity param=0.1585
  predicted class=No    expected loss=0.1602
    class counts:    215    41
   probabilities:  0.840 0.160
  left son=2 (204 obs) right son=3 (52 obs)
  Primary splits:
   Pressure3pm    < 1012   right, improve=13.420, (0 missing)
   Cloud3pm       < 7.5    left,  improve= 9.492, (0 missing)
   Pressure9am    < 1016   right, improve= 9.143, (0 missing)
   Sunshine       < 6.45   right, improve= 8.990, (2 missing)
   WindGustSpeed < 64      left,  improve= 7.339, (2 missing)
  Surrogate splits:
   Pressure9am    < 1013   right, agree=0.938, adj=0.692,...
   MinTemp        < 16.15 left,   agree=0.824, adj=0.135,...
   Temp9am        < 20.35 left,   agree=0.816, adj=0.096,...
   WindGustSpeed < 64      left,   agree=0.812, adj=0.077,...
   WindSpeed3pm   < 34     left,   agree=0.812, adj=0.077,...
```

We see that node number 1 has 256 observations to work with. It has a complexity parameter of 0.1585366, which is discussed in Section 11.5.

The next line identifies the default class for this node (No in this case) which corresponds to the class that occurs most frequently in the training dataset. With this class as the decision associated with this node, the error rate (or expected loss) is 16% (or 0.1601562).

The table that follows then reports the frequency of observations by the target variable. There are 215 observations with No for RainTomorrow (84%) and 41 with Yes (16%).

The remainder of the information relates to deciding how to split the node into two subsets. The resulting split has a left branch (labelled as node number 2) with 204 observations. The right branch (labelled as node number 3) has 52 observations.

The actual variable used to split the dataset into these two subsets is Pressure3pm, with the test being on the value 1011.9. Any observation with Pressure3pm < 1011.9 goes to the right branch, and so ≥ 1011.9 goes to the left. The measure (the improvement) associated with this split of the dataset is 13.42.

We then see a collection of alternative splits and their associated measures. Clearly, Pressure3pm offers the best improvement, with the nearest competitor offering an improvement of 9.49.

The surrogate splits that are then presented relate to the handling of missing values in the data. Consider the situation where we apply the model to new data but have an observation with **Pressure3pm** missing. We could instead use **Pressure9am**. The information here indicates that 93.8% of the observations in the split based on *Pressure9am* < 1013.3 are the same as that based on *Pressure3pm* < 1011.9. The **adj** value is an indication of what is gained by using this surrogate split over simply giving up at this node and assigning the majority decision to the new observation. Thus, in using **Pressure9am** we gain a 69% improvement by using the surrogate.

The other nodes are then listed in the summary. They include the same kind of information, and we see at the beginning of node number 2 here:

```
Node number 2: 204 observations, complexity param=0.07317
  predicted class=No    expected loss=0.07843
    class counts:    188    16
   probabilities: 0.922 0.078
  left son=4 (195 obs) right son=5 (9 obs)
  Primary splits:
    Cloud3pm    < 7.5   left,   improve=6.516, (0 missing)
    Sunshine    < 6.4   right,  improve=2.937, (2 missing)
    Cloud9am    < 7.5   left,   improve=2.795, (0 missing)
    Humidity3pm < 71    left,   improve=1.465, (0 missing)
    WindDir9am  splits as  RRRRR...LLLL, improve=1.391,...

...
```

Note how categoric variables are reported. **WindDir9am** has 16 levels:

```
> levels(weather$WindDir9am)

 [1] "N"    "NNE" "NE"  "ENE" "E"   "ESE" "SE"  "SSE" "S"
[10] "SSW" "SW"  "WSW" "W"   "WNW" "NW"  "NNW"
```

All possible binary combinations of levels will have been considered and the one reported above offers the best improvement. Here the first five levels (N to E) correspond to the right (R) branch and the remainder to the left (L) branch.

The leaf nodes of the decision tree (nodes 4, 5, 6, and 7) will have just the relevant information—thus no information on splits or surrogates. An example is node 7. The following text again comes from the output of summary():

```
. . .

Node number 7: 27 observations
  predicted class=Yes  expected loss=0.2593
    class counts:      7     20
   probabilities: 0.259 0.741
```

Node 7 is a leaf node that predicts Yes as the outcome. The error/loss is 7 out of 27, or 0.2593 or 26%, and the probability of Yes is 74%.

Miscellaneous Functions

We have covered above the main functions and commands in R for building and displaying a decision tree. **Rpart** and **rattle** also provide a collection of utility functions for exploring the model.

First, the where= component of the decision tree object records the leaf node of the decision tree in which each observation in the training dataset ends up:

```
> head(weatherRPART$model$where, 12)

335 343 105 302 233 188 266  49 236 252 163 256
  3   3   7   3   3   3   3   4   3   3   3   3
```

The plot() command and the related text() command will display a decision tree labelled appropriately:

```
> opar <- par(xpd=TRUE)
> plot(weatherRPART$model)
> text(weatherRPART$model)
> par(opar)
```

We notice that the default plot (Figure 11.4) looks different from the plot we obtain through Rattle. Rattle provides drawTreeNodes() as a variation of plot() based on draw.tree() from **maptree** (White, 2010). The plot here is a basic plot. The length of each line within the tree branches gives a visual indication of the error down that branch of the tree. The plot and text can be further tuned through addition arguments to the two commands. There are very many tuning options available,

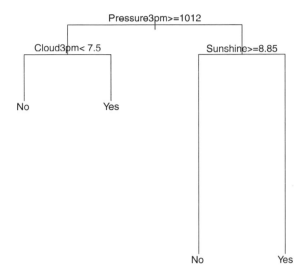

Figure 11.4: Typical R decision tree.

and they are listed in the manuals for the commands (?plot.rpart and ?text.rpart). The path.rpart() command is then a useful adjunct to plot():

```
> path.rpart(weatherRPART$model)
```

Running this command allows us to use the left mouse button to click on a node on the plot to list the path to that node. For example, clicking the left mouse button on the bottom right node results in:

```
node number: 7
  root
  Pressure3pm< 1012
  Sunshine< 8.85
```

Click on the middle or right mouse button to finish interacting with the plot.

11.5   Tuning Parameters

Any implementation of the decision tree algorithm provides a collection of parameters for tuning how the tree is built. The defaults in Rattle (based on **rpart**'s defaults) often provide a basically good tree. They are certainly a very good starting point and may be a satisfactory end point, too. However, tuning will be necessary where, for example, the target variable has very few examples of the particular class of interest or we would like to explore a number of alternative trees.

Whilst many tuning parameters are introduced here in some level of detail, the R documentation provides much more information. Use ?rpart to start exploring further. The rpart() function has two arguments for tuning the algorithm, each being a structure containing other options. They are control= and parms=. We use these as in the following example:

```
> evalq({
    control <- rpart.control(minsplit=10,
                             minbucket=5,
                             maxdepth=20,
                             usesurrogate=0,
                             maxsurrogate=0)
    model <- rpart(formula=form,
                   data=data[train, vars],
                   method="class",
                   parms=list(split="information"),
                   control=control)
}, weatherRPART)
```

We have already discussed the formula= and data= arguments. The remaining arguments are now discussed.

Modelling Method (method=)

The method= argument indicates the type of model to be built and is dependent on the target variable. For categoric targets, we generally build classification models, and so we use method="class". If the target is a numeric variable, then the argument would be method="anova" for an "analysis of variance," building a regression tree.

Splitting Function (split=)

The `split=` argument is used to choose between different splitting functions (measures). The argument appears within the `parms` argument of `rpart()`, which is built up as a named list. The `split="information"` directs `rpart` to use the information gain measure we introduced above. The default choice of `split="gini"` (in R, though Rattle's default is `"information"`) uses the Gini index of diversity. The choice makes no difference in this case, as we can verify by reviewing the output of the following two commands (though here we show just the one set of output):

```
> evalq({
    rpart(formula=form,
          data=data[train, vars],
          parms=list(split="information"))
  }, weatherRPART)

n= 256

node), split, n, loss, yval, (yprob)
      * denotes terminal node

1) root 256 41 No (0.83984 0.16016)
  2) Pressure3pm>=1012 204 16 No (0.92157 0.07843)
    4) Cloud3pm< 7.5 195 10 No (0.94872 0.05128) *
    5) Cloud3pm>=7.5 9  3 Yes (0.33333 0.66667) *
  3) Pressure3pm< 1012 52 25 No (0.51923 0.48077)
    6) Sunshine>=8.85 25  5 No (0.80000 0.20000) *
    7) Sunshine< 8.85 27  7 Yes (0.25926 0.74074) *
```

```
> evalq({
    rpart(formula=form,
          data=data[train, vars],
          parms=list(split="gini"))
  }, weatherRPART)
```

Minimum Split (`minsplit=`)

The `minsplit=` argument specifies the minimum number of observations that must exist at a node in the tree before it is considered for splitting. A node is not considered for splitting if it has fewer than `minsplit` observations. The `minsplit=` argument appears within the `control=` argument of `rpart()`. The default value of `minsplit=` is 20.

In the following example, we illustrate the boundary between splitting and not splitting the root node of our decision tree. This is often an issue in building a decision tree, and an inconvenience when all we obtain is a root node. Here the example shows that with a `minsplit=` of 53 the tree building will not proceed past the root node:

```
> evalq({
    rpart(formula=form,
          data=data[train, vars],
          control=rpart.control(minsplit=53))
  }, weatherRPART)
n= 256

node), split, n, loss, yval, (yprob)
      * denotes terminal node

1) root 256 41 No (0.8398 0.1602) *
```

Setting `minsplit=` to 52 results in a split on `Pressure3pm` (and further splitting) being considered and chosen, as we see in the code block below. Splitting on `Pressure3pm` splits the dataset into two datasets, one with 204 observations and the other with 52 observations. We can then see why, with `minsplit=` set to of 53, the tree building does not proceed past the root node.

Changing the value of `minsplit=` allows us to eliminate some computation, as nodes with a small number of observations will generally play less of a role in our models. Leaf nodes can still be constructed that have fewer observations than the `minsplit=`, as that is controlled by the `minbucket=` argument.

```
> evalq({
    rpart(formula=form,
          data=data[train, vars],
          control=rpart.control(minsplit=52))
  }, weatherRPART)

n= 256

node), split, n, loss, yval, (yprob)
      * denotes terminal node

1) root 256 41 No (0.83984 0.16016)
  2) Pressure3pm>=1012 204 16 No (0.92157 0.07843) *
  3) Pressure3pm< 1012 52 25 No (0.51923 0.48077)
    6) Sunshine>=8.85 25  5 No (0.80000 0.20000) *
    7) Sunshine< 8.85 27  7 Yes (0.25926 0.74074) *
```

Minimum Bucket Size (minbucket=)

The minbucket= argument is the minimum number of observations in
any leaf node. The default value is 7, or about one-third of the default
value of minsplit=. If either of these two arguments is specified but
not the other, then the default of the unspecified one is taken to be a
value such that this relationship holds (i.e., minbucket= is one-third of
minsplit=)

Once again we will see two examples of using minbucket=. The first
example limits the minimum bucket size to be 10, resulting in the same
model we obtained above. The second example reduces the limit down
to just 5 observations in the bucket. The result will generally be a larger
decision tree, since we are allowing leaf nodes with a smaller number of
observations to be considered, and hence the option to split a node into
smaller nodes will often be exercised by the tree building algorithm.

```
> ops <- options(digits=2)
> evalq({
    rpart(formula=form,
          data=data[train, vars],
          control=rpart.control(minbucket=10))
  }, weatherRPART)
n= 256

node), split, n, loss, yval, (yprob)
      * denotes terminal node

1) root 256 41 No (0.840 0.160)
  2) Pressure3pm>=1e+03 204 16 No (0.922 0.078) *
  3) Pressure3pm< 1e+03 52 25 No (0.519 0.481)
    6) Sunshine>=8.9 25  5 No (0.800 0.200) *
    7) Sunshine< 8.9 27  7 Yes (0.259 0.741) *
> evalq({
    rpart(formula=form,
          data=data[train, vars],
          control=rpart.control(minbucket=5))
  }, weatherRPART)
n= 256

node), split, n, loss, yval, (yprob)
      * denotes terminal node

 1) root 256 41 No (0.840 0.160)
   2) Pressure3pm>=1e+03 204 16 No (0.922 0.078)
     4) Cloud3pm< 7.5 195 10 No (0.949 0.051) *
     5) Cloud3pm>=7.5 9  3 Yes (0.333 0.667) *
   3) Pressure3pm< 1e+03 52 25 No (0.519 0.481)
     6) Sunshine>=8.9 25  5 No (0.800 0.200) *
     7) Sunshine< 8.9 27  7 Yes (0.259 0.741)
      14) Evaporation< 5.5 15  7 Yes (0.467 0.533)
         28) WindGustSpeed< 58 10  3 No (0.700 0.300) *
         29) WindGustSpeed>=58 5  0 Yes (0.000 1.000) *
      15) Evaporation>=5.5 12  0 Yes (0.000 1.000) *
> options(ops)
```

Note that changing the value of `minbucket=` can have an impact on the choice of variable for the split. This will occur when one choice with a higher improvement results in a node with too few observations, leading to another choice being taken to meet the minimum requirements for the number of observations in a split.

Whilst the default is to set `minbucket=` to be one-third of `minsplit=`, there is no requirement for `minbucket=` to be less than `minsplit=`. A node will always have at least `minbucket=` entities, and it will be considered for splitting if it has at least `minsplit=` observations and if on splitting each of its children has at least `minbucket=` observations.

Complexity Parameter (`cp=`)

The complexity parameter is used to control the size of the decision tree and to select an optimal tree size. The complexity parameter controls the process of pruning a decision tree. As we will discuss in Chapter 15, without pruning, a decision tree model can overfit the training data and then not perform very well on new data. In general, the more complex a model, the more likely it is to match the data on which it has been trained and the less likely it is to match new, previously unseen data.

On the other hand, decision tree models are very interpretable, and thus building a more complex tree (i.e., having many branches) is sometimes tempting (and useful). It can provide insights that we can then test statistically.

Using `cp=` governs the minimum "benefit" that must be gained at each split of the decision tree in order to make a split worthwhile. This therefore saves on computing time by eliminating splits that appear to add little value to the model. The default is 0.01. A value of 0 will build a "complete" decision tree to maximum depth depending on the values of `minplit=` and `minbucket=`. This is useful if we want to look at the values for CP for various tree sizes. We look for the number of splits where the sum of the xerror (cross-validation error relative to the root node error) and xstd is minimum (as discussed in Section 11.4). This is usually early in the list.

The `plotcp()` command is useful in visualising the progression of the CP values. In the following example,[1] we build a full decision tree with

---

[1]Note that the *cptable* may vary slightly between different deployments of R, particularly between 64 bit R, as here, and 32 bit R.

both `cp=` and `minbucket=` set to zero. We also show the CP table. The corresponding plot is shown in Figure 11.5.

```
> set.seed(41)
> evalq({
    control <- rpart.control(cp=0, minbucket=0)
    model <- rpart(formula=form,
                   data=data[train, vars],
                   control=control)
  }, weatherRPART)
> print(weatherRPART$model$cptable)

        CP nsplit rel error xerror   xstd
1 0.15854      0   1.00000 1.0000 0.1431
2 0.07317      2   0.68293 0.9024 0.1372
3 0.04878      3   0.60976 0.9024 0.1372
4 0.03659      7   0.41463 0.8780 0.1357
5 0.02439     10   0.29268 0.8780 0.1357
6 0.01829     13   0.21951 1.0488 0.1459
7 0.01220     21   0.02439 1.1463 0.1511
8 0.00000     23   0.00000 1.2439 0.1559

> plotcp(weatherRPART$model)
> grid()
```

The figure illustrates a typical behaviour of model building. As we proceed to build a complex model, the error rate (the y-axis) initially decreases. It then flattens out and, as the model becomes more complex, the error rate begins to again increase. We will want to choose a model where it has flattened out. Based on the principle of favouring simpler models, we might choose the first of the similarly performing bottom points and thus we might set `cp=` 0.1, for example.

As a script, we could automate the selection with the following:

```
> xerr <- weatherRPART$model$cptable[,"xerror"]
> minxerr <- which.min(xerr)
> mincp <- weatherRPART$model$cptable[minxerr, "CP"]
> weatherRPART$model.prune <- prune(weatherRPART$model,
                                    cp=mincp)
```

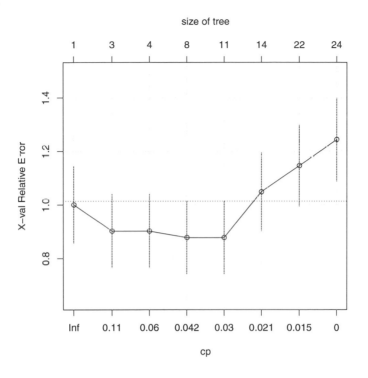

Figure 11.5: Error rate versus complexity/tree size.

Priors (prior=)

Sometimes the proportions of classes in a training set do not reflect their true proportions in the population. We can inform Rattle and R of the population proportions, and the resulting model will reflect them. All probabilities will be modified to reflect the prior probabilities of the classes rather than the actual proportions exhibited in the training dataset.

The priors can also be used to "boost" a particularly important class, by giving it a higher prior probability, although this might best be done through the loss matrix (Section 11.5).

In Rattle, the priors are expressed as a list of numbers that sum to 1. The list must be of the same length as the number of unique classes in the training dataset. An example for binary classification is 0.6,0.4. This translates into prior=c(0.6,0.4) for the call to rpart().

The following example illustrates how we might use the priors to

favour a particular target class that was otherwise not being predicted
by the resulting model (because the resulting model turns out to be only
a root node always predicting No). We begin by creating the dataset
object, consisting of the larger Australian weather dataset, *weatherAUS*:

```
> wausDS <- new.env()
> evalq({
    data <- weatherAUS
    nobs <- nrow(data)
    form <- formula(RainTomorrow ~ RainToday)
    target <- all.vars(form)[1]
    set.seed(42)
    train <- sample(nobs, 0.5*nobs)
  }, wausDS)
```

A decision tree model is then built and displayed:

```
> wausRPART <- new.env(parent=wausDS)
> evalq({
    model <- rpart(formula=form, data=data[train,])
    model
  }, wausRPART)
n=19509 (489 observations deleted due to missingness)

node), split, n, loss, yval, (yprob)
      * denotes terminal node

1) root 19509 4491 No (0.7698 0.2302) *
```

A table shows the proportion of observations assigned to each class in
the training dataset.

```
> evalq({
    freq <- table(data[train, target])
    round(100*freq/length(train), 2)
  }, wausRPART)

   No    Yes
75.66 22.74
```

Now we build a decision tree model but with different prior probabilities:

```
> evalq({
    model <- rpart(formula=form,
                    data=data[train,],
                    parm=list(prior=c(0.5, 0.5)))
    model
  }, wausRPART)

n=19509 (489 observations deleted due to missingness)

node), split, n, loss, yval, (yprob)
      * denotes terminal node

1) root 19509 9754 Yes (0.5000 0.5000)
  2) RainToday=No 15042 5098 No (0.6180 0.3820) *
  3) RainToday=Yes 4467 1509 Yes (0.2447 0.7553) *
```

The default priors when using `raprt()` without the `prior=` option are set to be the class proportions as found in the training dataset supplied.

Loss Matrix (`loss=`)

The loss matrix is used to weight different kinds of errors (or loss) differently. This refers to what are commonly known as false positives (or type I errors) and false negatives (or type II errors) when we talk about a two-class problem.

Often, one type of error is more significant than another type of error. In fraud, for example, a model that identifies too many false positives is probably better than a model that identifies too many false negatives (because we then miss too many real frauds). In medicine, a false positive means that we diagnose a healthy patient with a disease, whilst a false negative means that we diagnose an ill patient as being healthy.

The default loss for each of the true/false positives/negatives is 1— they are all of equal impact or loss. In the case of a rare, and under-represented class (like fraud) we might consider false negatives to be four or even ten times worse than a false positive. Thus, we communicate this to the algorithm so that it will work harder to build a model to find all of the positive cases.

The loss matrix records these relative weights for the two class case only. The following table illustrates the terminology (showing predicted versus observed):

$$
\begin{array}{ccc}
Pr\ v\ Ob & 0 & 1 \\
0 & TN & FN \\
1 & FP & TP
\end{array}
$$

Noting that we do not specify any weights in the loss matrix for the true positives (TP) and the true negatives (TN), we supply weights of 0 for them in the matrix. To specify the matrix in the Rattle interface, we supply a list of the form: $0, FN, FP, 0$.

In general, the loss matrix must have the same dimensions as the number of classes (i.e., the number of levels of the target variable) in the training dataset. For binary classification, we must supply four numbers with the diagonals as zeros.

An example is the string of numbers $0, 10, 1, 0$, which might be interpreted as saying that an actual 1 predicted as 0 (i.e., a false negative) is ten times more unwelcome than a false positive. This is used to construct, row-wise, the loss matrix which is passed through to rpart() as `loss=loss=matrix(c(0,10,1,0), byrow=TRUE, nrow=2)`.

The loss matrix is used to alter the priors, which will affect the choice of variable on which to split the dataset on at each node, giving more weight where appropriate.

Using the loss matrix is often indicated when we build a decision tree that ends up being just a single root node (often because the positive class represents less than 5% of the population—and so the most accurate model would predict everyone to be a negative).

Other Options

The rpart() function provides many other tuning parameters that are not exposed through the Rattle GUI. These include maxdepth= to limit the depth of a tree and maxcompete= to limit the number of competing alternative splits for each node that is retained in the resulting model.

A number of options relate to the handling of surrogates. As indicated above, surrogates in the model allow for the handling of missing values. The surrogatestyle= argument indicates how surrogates are given preference. The default is to prefer variables with fewer missing values in the training dataset, with the alternative being to sort them by the percentage correct over the number of nonmissing values.

The usesurrogate= argument controls how surrogates are made use of in the model. The default for the usesurrogate= argument is 2.

This is also set when **Rattle's Include Missing** check button is active. The behaviour here is to try each of the surrogates whenever the main variable has a missing value, but if all surrogates are also missing, then follow the path with the majority of cases. If `usesurrogate=` is set to 1, the behaviour is to try each of the surrogates whenever the main variable has a missing value, but if all surrogates are also missing, then go no further. When the argument is set to 0 (the case when **Rattle's Include Missing** check button is not active), the observation with a missing value for the main variable is not used any further in the tree building.

The `maxsurrogate=` argument simply limits the number of surrogates considered for each node.

11.6   Discussion

Decision trees have been around for a long time. They present a mechanism for structuring a series of questions. The next question to ask, at any time, is based on the answer to a previous question.

In data mining, we commonly identify decision trees as the knowledge representation scheme targeted by the family of techniques originating from ID3 within the machine learning community (Quinlan, 1986) and from CART within the statistics community. The original ID3 algorithm was extended to become the commercially available C4.5 software. This was made available together with a book by Quinlan (1993) that served as a guide to using the code.

Traditional decision tree algorithms can suffer from overfitting and can exhibit a bias towards selecting variables with many possible splits (i.e., categoric variables). The algorithms do not use any statistical significance concepts and thus, as noted by Mingers (1989), cannot distinguish between significant and insignificant improvements in the information measure. The use of a cross-validated relative error measure, as in the implementation in `rpart()` does guard against overfitting.

Hothorn et al. (2006) introduced an improvement to the approach presented here for building a decision tree, called conditional inference trees. Rattle offers the choice of traditional and conditional algorithms. Conditional inference trees address overfitting and variable selection biases by using a conditional distribution to measure the association between the output and the input variables. They take into account distributional properties.

Conditional inference trees can be built using `ctree()` from **party** (Hothorn et al., 2006). Within Rattle, we can choose the Conditional option to build a conditional inference tree. From the command line, we would use the following call to `ctree()`:

```
> library(party)
> weatherCTREE <- new.env(parent=weatherDS)
> evalq({
    model <- ctree(formula=form, data=data[train, vars])
  }, weatherCTREE)
```

We can review just lines 8 to 17 of the resulting output, which is the tree itself:

```
> cat(paste(capture.output(weatherCTREE$model)[8:17],
            collapse="\n"))

1) Pressure3pm <= 1012; criterion = 1, statistic = 39.281
  2) Sunshine <= 8.8; criterion = 0.99, statistic = 12.099
    3)*  weights = 27
  2) Sunshine > 8.8
    4)*  weights = 25
1) Pressure3pm > 1012
  5) Cloud3pm <= 7; criterion = 1, statistic = 20.825
    6)*  weights = 195
  5) Cloud3pm > 7
    7)*  weights = 9
```

A plot of the tree is presented in Figure 11.6. The plot is quite informative and primarily self-explanatory. Node 3, for example, predicts rain relatively accurately, whilst node 6 describes conditions under which there is almost never any rain on the following day.

11.7 Summary

Decision tree algorithms handle mixed types of variables and missing values, and are robust to outliers and monotonic transformations of the input and to irrelevant inputs. The predictive power of decision trees tends to be poorer than for other techniques that we will introduce. However, the algorithm is generally straightforward, and the resulting

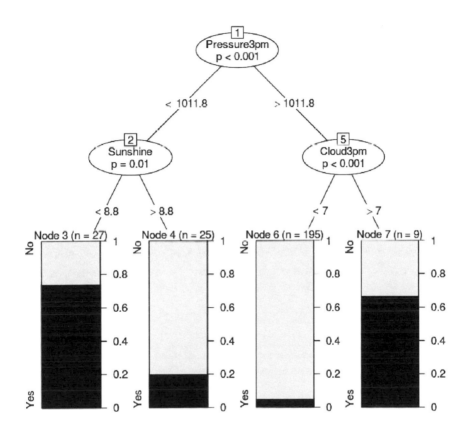

Figure 11.6: A conditional inference tree.

models are generally easily interpretable. This last characteristic has made decision tree induction very popular for over 30 years.

This chapter has introduced the basic concept of representing knowledge as a decision tree and presented a measure for choosing a good decision tree and an algorithm for building one.

11.8   Command Summary

This chapter has referenced the following R packages, commands, functions, and datasets:

`ctree()`	function	Build a conditional inference tree.
`draw.tree()`	command	Enhanced graphic decision tree.
maptree	package	Provides `draw.tree()`.
party	package	Conditional inference trees.
`path.rpart()`	function	Identify paths through decision tree.
`plot()`	command	Graphic display of the tree.
`plotcp()`	command	Plot complexity parameter.
`print()`	command	Textual version of the decision tree.
`printcp()`	command	Complexity parameter table.
rattle	package	The *weather* dataset and GUI.
`rpart()`	function	Build a decision tree predictive model.
rpart	package	Provides decision tree functions.
rpart.control	function	Organise `rpart` control arguments.
`set.seed()`	function	Initiate random seed number sequence.
`summary()`	command	Summary of the tree building process.
`text()`	command	Add labels to decision tree graphic.
weather	dataset	Sample dataset from **rattle**.

Chapter 12

Random Forests

Building a single decision tree provides a simple model of the world, but it is often too simple or too specific. Over many years of experience in data mining, it has become clear that many models working together are better than one model doing
it all. We have now become familiar with the idea of combining multiple models (like decision trees) into a single ensemble of models (to build a forest of trees).

Compare this to how we might bring together panels of experts to ponder an issue and to then come up with a consensus decision. Governments, industry, and universities all manage their business processes in this way. It can often result in better decisions compared to simply relying on the expertise of a single authority on a topic.

The idea of building multiple trees arose early on with the development of the multiple inductive learning (MIL) algorithm (Williams, 1987, 1988). In building a single decision tree, it was noted that often there was very little difference in choosing between alternative variables. For example, two or more variables might not be distinguishable in terms of their ability to partition the data into more homogeneous datasets. The MIL algorithm builds all "equally" good models and then combines them into one model, resulting in a better overall model.

Today we see a number of algorithms generating ensembles, including boosting, bagging, and random forests. In this chapter, we introduce the random forest algorithm, which builds hundreds of decision trees and combines them into a single model.

12.1 Overview

The random forest algorithm tends to produce quite accurate models because the ensemble reduces the instability that we can observe when we build single decision trees. This can often be illustrated simply by removing a very small number of observations from the training dataset, to see quite a change in the resulting decision tree.

The random forest algorithm (and other ensemble algorithms) tends to be much more robust to changes in the data. Hence,it is very robust to noise (i.e., variables that have little relationship to the target variable). Being robust to noise means that small changes in the training dataset will have little, if any, impact on the final decisions made by the resulting model. Random forest models are generally very competitive with nonlinear classifiers such as artificial neural nets and support vector machines.

Random forests handle underrepresented classification tasks quite well. This is where, in the binary classification task, one class has very few (e.g., 5% or fewer) observations compared with the other class.

By building each decision tree to its maximal depth, as the random forest algorithm does (by not pruning the individual decision trees), we can end up with a model that is less biased. Each individual tree will overfit the data, but this is outweighed by the multiple trees using different variables and (over) fitting the data differently.

The randomness used by a random forest algorithm is in the selection of both observations and variables. It is this randomness that delivers considerable robustness to noise, outliers, and overfitting, when compared with a single-tree classifier.

The randomness also delivers substantial computational efficiencies. In building a single decision tree, the model builder may select a random subset of the observations available in the training dataset. Also, at each node in the process of building the decision tree, only a small fraction of all of the available variables are considered when determining how to best partition the dataset. This substantially reduces the computational requirement.

In the area of genetic marker selection and microarray data within bioinformatics, for example, random forests have been found to be particularly well suited. They perform well even when many of the input variables have little bearing on the target variable (i.e., they are noise variables). Random forests are also suitable when there are very many input variables and not so many observations.

In summary, a random forest model is a good choice for model building for a number of reasons. Often, very little preprocessing of the data needs to be performed, as the data does not need to be normalised and the approach is resilient to outliers. The need for variable selection is avoided because the algorithm effectively does its own. Because many trees are built using two levels of randomness (observations and variables), each tree is effectively an independent model and the resulting model tends not to overfit to the training dataset.

12.2   Knowledge Representation

The random forest algorithm is commonly presented in terms of decision trees as the primary form for representing knowledge. However, the random forest algorithm can be thought of as a meta-algorithm. It describes an approach to building models where the actual model builder could be a decision tree algorithm, a regression algorithm, or any one of many other kinds of model building algorithms. The general concepts apply to any of these approaches. We will stay with decision trees as the underlying model builder for our purposes here.

In any ensemble approach, the key extension to the knowledge representation is in the way that we combine the decisions that are made by the individual "experts" or models. Various approaches have been considered over the years. Many come from the knowledge-based and expert systems communities, which often need to consider the issue of combining expert knowledge from multiple experts. Approaches to aggregating decisions into one final decision include simple majority rules and a weighted score where the weights correspond to the quality of the expertise (e.g., the measured accuracy of the individual tree).

The random forest algorithms will often build from 100 to 500 trees. In deploying the model, the decisions made by each of the trees are combined by treating all trees as equals. The final decision of the ensemble will be the decision of the majority of the constituent trees. If 80 out of 100 trees in the random forest say that it will rain tomorrow, then we will go with that decision and take the appropriate action for rain. Even if 51 of the 100 trees say that it will rain, we might go with that, although perhaps with less certainty. In the context of regression rather than classification, the result is the average value over the ensemble of regression trees.

12.3   Algorithm

Chapter 11 covered the building of an individual tree, and the same algorithm can be used for building one or 500 trees. It is how the training set is selected and how the variables to use in the modelling are chosen that differs between the trees built for a random forest.

Sampling the Dataset

The random forest algorithm builds multiple decision trees, using a concept called *bagging*, to introduce random sampling into the whole process. Bagging is the idea of collecting a random sample of observations into a bag (though the term itself is an abbreviation of *bootstrap aggregation*). Multiple bags are made up of randomly selected observations obtained from the original observations from the training dataset.

The selection in bagging is made with replacement, meaning that a single observation has a chance of appearing multiple times within a single bag. The sample size is often the same as for the full dataset, and so in general about two-thirds of the observations will be included in the bag (with repeats) and one-third will be left out. Each bag of observations is then used as the training dataset for building a decision tree (and those left out can be used as an independent sample for performance evaluation purposes).

Sampling the Variables

A second key element of randomness relates to the choice of variables for partitioning the dataset. At each step in building a single decision node (i.e., at each split point of the tree), a random, and usually small, set of variables is chosen. Only these variables are considered when choosing a split point. For each node in building a decision tree, a different random set of variables is considered.

Randomness

By randomly sampling both the data and the variables, we introduce decision trees that purposefully have different performance behaviours for different subsets of the data. It is this variation that allows us to consider an ensemble of such trees as representing a team of experts with differing expertise working together to deliver a "better" answer.

Sampling also offers another significant advantage—computational efficiency. By considering only a small fraction of the total number of variables available, whilst considering split points, the amount of computation required is significantly reduced.

In building each decision tree, the random forest algorithm generally will not perform any pruning of the decision tree. When building a single decision tree, it was noted in Chapter 11 that pruning was necessary to avoid overfitting the data. Overfitted models tend not to perform well on new data. However, a random forest of overfitted trees can deliver a very good model that performs well on new data.

Ensemble Scoring

In deploying the multiple decision trees as a single model, each tree has equal weight in the final decision-making process. A simple majority might dictate the outcome. Thus, if 300 of 500 decision trees all predict that it will rain tomorrow, then we might be inclined to expect there to be rain tomorrow. If only 100 trees of the 500 predict rain tomorrow, then we might not expect rain.

12.4   Tutorial Example

Our task is again to predict the likelihood of rain tomorrow given today's weather conditions. We illustrate this using Rattle and directly through R. In both cases, **randomForest** (Liaw and Wiener, 2002) is used. This package provides direct access to the original implementation of the random forest algorithm by its authors.

Building a Model using Rattle

Rattle's Model tab provides the Forest option to build a forest of decision trees. Figure 12.1 displays the graphical interface to the options for building a random forest with the default values and also shows the top part of the results from building the random forest shown in the Textview area.

We now step through the output of the text view line by line. The first few lines note the number of observations used to build the model and then an indication that missing values in the training dataset are being imputed. If missing value imputation is not enabled, then the

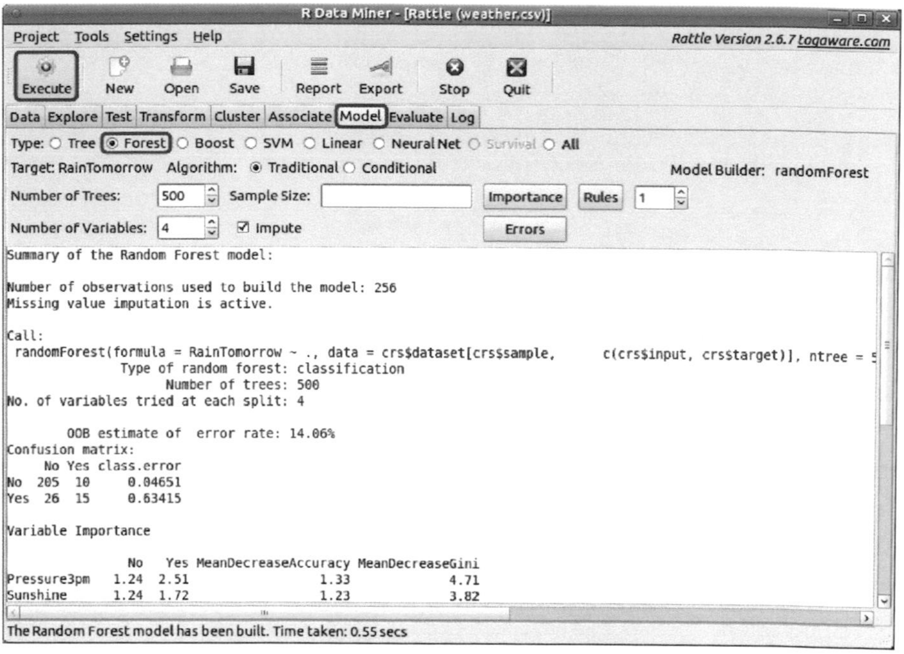

Figure 12.1: Building a random forest predictive model.

number of observations may be less than the number available, as the default is to drop observations containing missing values.

```
Summary of the Random Forest model:

Number of observations used to build the model: 256
Missing value imputation is active.
```

The next few lines record the actual function command line call that Rattle generated and passed onto R to be evaluated:

```
Call:
 randomForest(formula = RainTomorrow ~ .,
      data = crs$dataset[crs$sample,
                         c(crs$input, crs$target)],
      ntree = 500, mtry = 4, importance = TRUE,
      replace = FALSE, na.action = na.roughfix)
```

A more detailed dissection of the function call is presented later, but in brief, 500 trees were asked for (`ntree=`) and just four variables were considered for the split point for each node (`mtry=`). An indication of the importance of variables is maintained (`importance=`), and any observations with missing values will have those values imputed (`na.action=`).

The next few lines summarise some of the same information in a more accessible form. Note that, due to numerical differences, specific results may vary slightly between 32 bit and 64 bit deployments of R. The following was performed on a 64 bit deployment of R:

```
            Type of random forest: classification
                  Number of trees: 500
No. of variables tried at each split: 4
```

Performance Evaluation

Next comes an indication of the performance of the resulting model. The out-of-bag (OOB) estimate of the error rate is calculated using the observations that are not included in the "bag"—the "bag" is the subset of the training dataset used for building the decision tree, hence the "out-of-bag" terminology.

This "unbiased" estimate of error suggests that when the resulting model is applied to new observations, the answers will be in error 14.06% of the time. That is, it is 85.94% accurate, which is a reasonably good model.

```
      OOB estimate of  error rate: 14.06%
```

This overall measure of accuracy is then followed by a confusion matrix that records the disagreement between the final model's predictions and the actual outcomes of the training observations. The actual observations are the rows of this table, whilst the columns correspond to what the model predicts for an observation and the cells count the number of observations in each category. That is, the model predicts Yes and the observation was No for 26 observations.

```
Confusion matrix:
      No Yes class.error
No   205  10     0.04651
Yes   26  15     0.63415
```

We see that the model and the training dataset agree that it won't rain for 205 of the observations. They agree that it will rain for 15 of the observations. However, there are 26 days for which the model predicts that it does not rain the following day and yet it does rain. Similarly, the model predicts that it will rain the following day for ten of the observations when in fact it does not rain.

The overall class errors, also calculated from the out-of-bag data, are included in the table. The model is wrong in predicting rain when there is none in only 63.41% of the observations when there is no rain. This is contrasted with the 4.65% error rate in predicting that it does rain tomorrow.

Underrepresented Classes

The acceptability of such errors (false positives versus false negatives) depends on many factors. Predicting that it will rain tomorrow and getting it wrong (false positive) might be an inconvenience in terms of carrying an umbrella around for the day. However, predicting that it won't rain and not being prepared for it (false negative) could result in a soggy dash for cover. The 63.41% error rate in predicting that it does not rain might be a concern.

One approach with random forests in addressing the "seriousness" associated with the false negatives might be to adjust the balance between the underrepresented class (66 observations have RainTomorrow as Yes) and the overrepresented class (300 observations have RainTomorrow as No). In the training dataset the observations are 41 and 215, respectively (after removing any observations with missing values).

We can use the Sample Size option to encourage the algorithm to be more aggressive in predicting that it will rain tomorrow. We will balance up the sampling so that equal numbers of observations with Yes and No are chosen. Specifying a value of 35,35 for the sample size will do this. The confusion matrix for the resulting random forest is:

```
          OOB estimate of  error rate: 28.52%

Confusion matrix:

      No Yes class.error
No   147  68      0.3163
Yes    5  36      0.1220
```

The error rate for when it does rain tomorrow is now 12.2%, and now we'll get wet 5 days out of 41 when it does rain, which is better than 26 days out of 41 days on which we'll end up getting wet.

The price we pay for this increased accuracy in predicting when it rains, is that we now have more days predicted as raining when in fact it does not rain. The "business problem" here indicates that carrying an umbrella with us unnecessarily is less of a burden than getting wet when it rains and we don't have our umbrella. We are also assuming that we don't want to carry our umbrella all the time.

Variable Importance

One of the problems with a random forest, compared with a single decision tree, is that it becomes quite a bit more difficult to readily understand the discovered knowledge—there are 500 trees here to try to understand. One way to get an idea of the knowledge being discovered is to consider the importance of the variables, as emerges from their use in the building of the 500 decision trees.

A variable importance table is the next piece of information that appears in the text view (we reformat it here to fit the limits of the page):

```
Variable Importance

                 No    Yes   Accu   Gini
Pressure3pm    1.24   2.51   1.33   4.71
Sunshine       1.24   1.72   1.23   3.82
Cloud3pm       1.13   1.90   1.16   3.19
WindGustSpeed  0.99   0.97   0.91   3.58
Pressure9am    1.03  -0.11   0.87   2.89
Temp3pm        0.83  -0.50   0.71   1.50
Humidity3pm    0.79   0.04   0.65   2.27
MaxTemp        0.61  -0.10   0.55   1.73
Temp9am        0.52   0.20   0.50   1.50
WindSpeed9am   0.56   0.12   0.46   1.39
```

The table lists each input variable and then four measures of importance for each variable. Higher values indicate that the variable is relatively more important. The table is sorted by the Accuracy measure of importance.

A naïve approach to measuring variable importance is to count the number of times the variable appears in the ensemble of decision trees. This is a rather blunt measure as, for example, variables can appear at different levels within a tree and thus have different levels of importance. Most measures thus incorporate some measure of the improvement made to the tree by each variable.

The third importance measure is a scaled average of the prediction accuracy of each variable. The calculation is based on a process of randomly permuting the values of a variable across the observations and measuring the impact on the predictive accuracy of the resulting tree. The larger the impact then the more important the variable is. Thus this measure reports the mean decrease in the accuracy of the model. The actual magnitude of the measure is not so relevant as the relative positioning of variables by the measure.

The final measure of importance is the total decrease in a decision tree node's impurity (the splitting criterion) when splitting on a variable. The splitting criterion used is the Gini index. This is measured for a variable over all trees giving a measure of the mean decrease in the Gini index of diversity relating to the variable.

The Importance button displays a visual plot of the accuracy and the Gini importance measures, as shown in Figure 12.2, and is more effective in illustrating the relative importance of the variables. Clearly, Pressure3pm is the most important variable, and then Sunshine. The accuracy measure then lists Cloud3pm and the next most important. This is consistent with the decision tree we built in Chapter 11. What we did not learn in building the decision tree is that Pressure9am is also quite important, and that the remainder are less so, at least according to the accuracy measure.

We also notice that the categoric variables (like the wind direction variables WindGustDir, WindDir9am, and WindDir3pm) have a higher importance according to the Gini measure than with the accuracy measure. This bias towards categoric variables with many categories, exhibited in the Gini measure, is discussed further in Section 12.6. It is noteworthy that this bias will mislead us about the importance of these categoric variables.

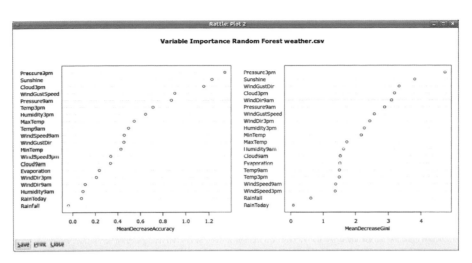

Figure 12.2: Two measures of variable importance as calculated by the random forest algorithm.

Time Taken

The tail of the textview provides information on how long it took to build the random forest of 500 trees. Note that even though we are building so many decision trees, the time taken is still less than 1 second.

Tuning Options

The Rattle interface provides a choice of Algorithm for building the random forest. The Traditional option is chosen by default, and that is what we have presented here. The Conditional option uses a more recent conditional inference tree algorithm for building the decision trees. This is explained in more detail in Section 12.6. A small number of other tuning options are also provided, and they are discussed in some detail in Section 12.5.

Error Plots

A useful diagnostic tool is the error plot, obtained with a click of the Error button. Figure 12.3 shows the resulting error plot for our random forest model.

The plot reports the accuracy of the forest of trees (in terms of error rate on the y-axis) against the number of trees that have been included

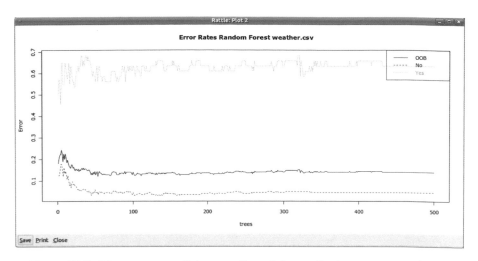

Figure 12.3: The error rate of the overall model gnerally decreases as each new tree is added to the ensemble.

in the forest (the x-axis). The key point we take from this plot is that after some number of trees there is actually very little that changes by adding further trees to the forest. From Figure 12.3 it would appear that going beyond about 20 trees in the forest adds very little value, when considering the out-of-bag (OOB) error rate.

The two other plots show the changes in error rate associated with the predictions of the model (here we have two classes predicted and so two additional lines). We also take these into account when deciding how many trees to add to the forest.

Conversion to Rules

Another button available with the **Forest** option is the **Rules** button, with an associated text entry box. Clicking this button will convert the specified tree into a set of rules. If the tree specified is 0 (rather than, for example, the default 1), then all trees will be converted to rules. Be careful, though, as that could take a very long time for 500 trees and 20 or more rules per tree (10,000 rules). The first two rules from tree 1 of the random forest are shown in the following code block.

```
Random Forest Model 1

Tree 1 Rule 1 Node 28 Decision No

1: Sunshine <= 6.45
2: Cloud9am <= 7.5
3: WindGustSpeed <= 43.5
4: Humidity3pm <= 36.5
5: MaxTemp <= 22.45

Tree 1 Rule 2 Node 29 Decision Yes

1: Sunshine <= 6.45
2: Cloud9am <= 7.5
3: WindGustSpeed <= 43.5
4: Humidity3pm <= 36.5
5: MaxTemp > 22.45
```

Building a Model Using R

As usual, we will create a container into which we place the relevant information for the modelling. We set up some useful variables within the container (using `evalq()`) as well as constructing the training and test datasets based on a random sample of 70% of the observations, including only those columns (i.e., dataset variables) that are not identified as being ignored (which is a list of negative indices, and thus indicates which columns not to include).

```
> library(rattle)
> weatherDS <- new.env()
> evalq({
    data <- na.omit(weather)
    nobs <- nrow(data)
    form <- formula(RainTomorrow ~ .)
    target <- all.vars(form)[1]
    vars <- -grep('^(Date|Location|RISK_)', names(data))
    set.seed(42)
    train <- sample(nobs, 0.7*nobs)
  }, weatherDS)
```

Considering the formula, the variable `RainTomorrow` is the target, with all remaining variables (~ .) from the provided dataset as the input variables.

Next we build the random forest. We first generate our training dataset as a random sample of 70% of the supplied dataset, noting that we reset the random number generator's seed back to a known number for repeatability. The data itself consists of the observations contained in the training dataset.

```
> library(randomForest)
> weatherRF <- new.env(parent=weatherDS)
> evalq({
    model <- randomForest(formula=form,
                          data=data[train, vars],
                          ntree=500, mtry=4,
                          importance=TRUE,
                          localImp=TRUE,
                          na.action=na.roughfix,
                          replace=FALSE)
  }, weatherRF)
```

The remaining arguments to the function are explained in Section 12.5.

Exploring the Model

The `model` object itself contains quite a lot of information about the model that has been built. The `str()` command gives the definitive list of all the components available within the object. An explanation is also available through the help page for `randomForest()`:

```
> str(weatherRF$model)
> ?randomForest
```

We consider some of the information stored within the object here.

The `predicted` component contains the values predicted for each observation in the training dataset based on the out-of-bag samples. If an observation is never in an out-of-bag sample then the prediction will be reported as `NA`. Here we show just the first ten predictions:

```
> head(weatherRF$model$predicted, 10)
```

```
336 342   94 304 227 173 265   44 230 245
 No  No   No  No  No  No  No   No  No  No
Levels: No Yes
```

The `importance` component records the information related to measures of variable importance as discussed in detail in Section 12.4, page 253. The information is reported for four measures (columns).

```
> head(weatherRF$model$importance)
```

	No	Yes	MeanDecreaseAccuracy
MinTemp	0.0031712	0.0056410	0.0036905
MaxTemp	0.0002405	0.0003834	0.0077143
Rainfall	0.0014129	-0.0033066	0.0005476
Evaporation	0.0006489	-0.0040790	-0.0002857
Sunshine	0.0211487	0.0445901	0.0251667
WindGustDir	0.0020603	0.0028510	0.0021905

	MeanDecreaseGini
MinTemp	2.2542
MaxTemp	1.8281
Rainfall	0.7377
Evaporation	1.3721
Sunshine	3.9320
WindGustDir	1.2739

The importance of each variable in predicting the outcome for each observation in the training dataset can also be available in the model object. This is accessible through the `localImp` component:

```
> head(weatherRF$model$localImp)[,1:4]
```

	336	342	94	304
MinTemp	0.011834	0.016575	0.021053	0.00000
MaxTemp	0.000000	0.005525	0.010526	-0.07143
Rainfall	0.005917	-0.005525	0.005263	0.00000
Evaporation	0.000000	0.000000	0.000000	-0.02976
Sunshine	0.035503	0.038674	-0.010526	0.03571
WindGustDir	0.005917	-0.005525	0.005263	0.04762

The error rate data is stored as the `err.rate` component. This can be accessed from the `model` object as we see in the following code block:

```
> weatherRF$model$err.rate
```

In **Rattle**, we saw an error plot that showed the change in error rate as more trees are added to the forest. We can obtain the actual data behind the plot quite easily:

```
> round(head(weatherRF$model$err.rate, 15), 4)

           OOB      No     Yes
 [1,]  0.2738  0.2143  0.5714
 [2,]  0.2701  0.2261  0.5000
 [3,]  0.2560  0.2340  0.3704
 [4,]  0.2273  0.1728  0.4722
 [5,]  0.2067  0.1361  0.5128
 [6,]  0.1872  0.1061  0.5500
 [7,]  0.1570  0.0984  0.4250
 [8,]  0.1689  0.1081  0.4500
 [9,]  0.1404  0.0691  0.4750
[10,]  0.1223  0.0529  0.4500
[11,]  0.1223  0.0582  0.4250
[12,]  0.1048  0.0317  0.4500
[13,]  0.1310  0.0582  0.4750
[14,]  0.1354  0.0529  0.5250
[15,]  0.1223  0.0476  0.4750
```

Here we see that the OOB estimate decreases quickly and then starts to flatten out. We can find the minimum quite simply, together with a list of the indexes where each minimum occurs:

```
> evalq({
    min.err <- min(data.frame(model$err.rate)["OOB"])
    min.err.idx <- which(data.frame(model$err.rate)["OOB"]
                     == min.err)
  }, weatherRF)
```

The actual minimum value together with the indexes can be listed:

```
> weatherRF$min.err

[1] 0.1048

> weatherRF$min.err.idx

[1] 12 45 49 50 51
```

We can then list the actual models where the minimum occurs:

```
> weatherRF$model$err.rate[weatherRF$min.err.idx,]

        OOB       No    Yes
[1,]  0.1048  0.03175  0.450
[2,]  0.1048  0.02116  0.500
[3,]  0.1048  0.01587  0.525
[4,]  0.1048  0.01587  0.525
[5,]  0.1048  0.01587  0.525
```

We might thus decide that 12 (the first instance of the minimum OOB estimate) is a good number of trees to have in the forest.

Another interesting component is **votes**, which records the number of trees that vote No and Yes within the ensemble for a particular observation.

```
> head(weatherRF$model$votes)

          No       Yes
336  0.9467  0.053254
342  0.9779  0.022099
94   0.8263  0.173684
304  0.8690  0.130952
227  0.9943  0.005682
173  0.9950  0.005025
```

The numbers are reported as proportions and so add up to 1 for each observation, as we can confirm:

```
> head(apply(weatherRF$model$votes, 1, sum))

336 342   94 304 227 173
  1   1    1   1   1   1
```

12.5 Tuning Parameters

Rattle provides access to just a few basic tuning options (Figure 12.1) for the random forest algorithm. The user interface allows the number of trees, the number of variables, and the sample size to be specified. As is generally the case with Rattle, the defaults are a good starting point!

These result in 500 trees being built, choosing from the square root of the number of variables available for each node, and no sampling of the training dataset to balance the classes.

In Figure 12.1, we see that the number of variables has automatically been set to 4 for the *weather* dataset, which has 20 input variables. The user interface options correspond to the function arguments `ntree=`, `ntry=`, and `sampsize=`. Rattle also sets `importance=` to TRUE, `replace=` to FALSE, and `na.action=` to `na.roughfix()`.

A call to `randomForest()` including all arguments covered here will look like:

```
> evalq({
    model <- randomForest(formula=form,
                          data=data[train, vars],
                          ntree=500,
                          mtry=4,
                          replace=FALSE,
                          sampsize=.632*nobs,
                          importance=TRUE,
                          localImp=FALSE,
                          na.action=na.roughfix)
  }, weatherRF)
```

Number of Trees `ntree=`

This specifies how many trees are to be built to populate the random forest. The default value is 500, and a common recommendation is that a minimum of 100 trees be built. The performance of the resulting random forest model tends not to degrade as the number of trees increases, though computationally it will take longer and will be more complex to use when scoring, and often there is little to gain from adding too many trees to a forest. The error matrix and error plot provide a guide to a good number of trees to include in a forest. See Section 12.4 for examples.

Number of Variables `ntry=`

The number of variables to consider for splitting at every node is specified by `ntry=`. This many variables will be randomly selected from all of those available each time we look to partition a dataset in the process of building the decision tree. The general default value is the square root of the total number of variables available, for classification tasks and one-third of the number of available variables for regression.

If there are many noise variables (i.e., variables that play little or no role in predicting the outcome), then we might consider increasing the number of variables considered at each node to ensure we have some relevant variables to choose from.

Sample Size `sampsize=`

The sample size argument can be used to force the algorithm to select a smaller sample size than the default or to sample the observations differently based on the output variable values (for classification tasks). For example, if our training dataset contains 5,000 observations for which it does not rain tomorrow and only 500 for which it does rain tomorrow, we can specify the sample size as `400,400`, for example, to have equal weight on both outcomes. This provides a mechanism for effectively setting the prior probabilities. See Section 12.4 for an example of doing this in Rattle.

Variable Importance `importance=`

The importance argument allows us to review the importance of each variable in determining the outcome. Two importance measures are calculated in addition to importance of the variable in relation to each outcome in a classification task. These have been described in Section 12.4, and issues with the measures are discussed in Section 12.6.

Sampling with Replacement `replace=`

By default, the sampling is performed when the training observations are sampled for building a particular tree within the forest samples with replacement. This means that any particular observation might appear multiple times within the sample, and thus some observations get over-represented in some datasets. This is a feature of the approach. The `replace=` argument set to `FALSE` will perform sampling without replacement.

Handling Missing Values `na.action=`

The implementation of the `randomForest()` algorithm does not directly handle missing values. A common approach on finding missing values is simply to ignore the observation with missing values by specifying `na.omit` as the value of `na.action=`. For some data, this could actually end up removing all observations from the training dataset. Another quick option is to replace missing values with the median (for numeric data) or the most frequent value (for categoric data) using `na.roughfix`.

12.6   Discussion

Brief History and Alternative Approaches

The concept of an ensemble of experts was something that the knowledge based and expert systems research communities were exploring in the 1980's. Some early work on building and combining multiple decision trees was undertaken at that time (Williams, 1988). Multiple decision trees were built by choosing different variables at nodes where the choice of variable was not clear. The resulting ensemble was found to produce a better predictive model.

Ho (1995, 1998) then developed the concept of randomly sampling variables to build the ensemble of decision trees. Half of the available variables were randomly chosen for building each of the decision tree. She noted that as more trees were added to the ensemble, the predictive performance increased, mostly monotonically.

Breiman (2001) built on the idea of randomly sampling variables by introducing random sampling of variables at each node as the decision tree is built. He also added the concept of bagging (Breiman, 1996) where different random samples of the dataset are chosen as the training dataset for each tree. His algorithm is in common use today, and his actual implementation can be accessed within R through **randomForest**.

In some situations we will have available a huge number of variables to choose from. Often only a small proportion of the available variables will have some influence on the target variable. By randomly selecting a small proportion of the available variables we will often miss the more relevant variables in building our trees.

An approach to address this situation introduces a weighted variable selection scheme to implement an enriched random forest (Amaratunga et al., 2008). Weights can be based on the q-value, derived from the p-value for a two-sample t-test. We test for a group mean effect of a variable, testing how well the variable can separate the values of the binary target variable. The resulting weights then bias the random selection of variables toward those that have more influence on the target variable.

An extension to this method allows it to work when the target variable has more than two values. In that case we can use a chi-square or information gain measure. The approach can be shown to produce considerably more accuracte models, by ensuring each decision tree has a high degree of independence from the other trees and by weighting the sampling of the variables to ensure important variables are selected for each tree.

Using Other Random Forests

The `randomForest()` function can also be applied to regression tasks, survival analysis, and unsupervised clustering (Shi and Horvath, 2006).

Limitation on Categories

An issue with the implementation of random forests in R is that it can not handle categoric data with more than 32 categoric values. Statistical concerns also suggest that categoric variables with more than 32 categories don't make a lot of sense, and thus little effort has been made in the R package to rectify the issue.

Importance Measures

We introduced the idea of measures of variable importance in building a model in Section 12.4. There we looked at the mean decrease in accuracy and the mean decrease in the Gini index as two measures calculated whilst the trees of the random forest are being built.

These variable importance measures provided by `randomForest()` have been found to be unreliable under certain conditions. The issue particularly arises where there is a mix of numeric and categoric variables or the numeric variables have quite different scales (e.g., Age versus Income), or then categoric variables have very different numbers of categories (Strobl et al., 2007). Less important variables can end up having too high an importance according to the measures used, and thus we will be misled into believing the measures provided. Indeed, the Gini measure can be quite biased, so that categorics with many categories obtain a higher importance.

The `cforest()` function of **party** (Hothorn et al., 2006) provides an improved importance measure. This newer measure can be applied to any dataset, using subsampling without replacement, to give a more reliable measure of variable importance. A key aspect is that rather than sampling the data with replacement to obtain a same size sample, a random subsample is used.

Underneath, `cforest()` builds conditional decision trees by using the `ctree()` function discussed in Chapter 11. In the following code block we first load **party** into the library and we create a new data structure to store our forest object, attaching the *weather* dataset to the object.

```
> library(party)
> weatherCFOREST <- new.env(parent=weatherDS)
```

Now we can build the model itself with a call to `cforest()`:

```
> evalq({
    model <- cforest(form,
                     data=data[vars],
                     controls=cforest_unbiased(ntree=50,
                       mtry=4))
  }, weatherCFOREST)
```

We could now explore the resulting forest, but here we will simply list the top few most important variables, according to the measure used by **party**:

```
> evalq({
    varimp <- as.data.frame(sort(varimp(model),
                                 decreasing=TRUE))
    names(varimp) <- "Importance"
    head(round(varimp, 4), 3)
  }, weatherCFOREST)

            Importance
Pressure3pm     0.0212
Sunshine        0.0163
Cloud3pm        0.0150
```

12.7   Summary

A random forest is an ensemble (i.e., a collection) of unpruned decision trees. Random forests are often used when we have very large training datasets and a very large number of input variables (hundreds or even thousands of input variables). A random forest model is typically made up of tens or hundreds of decision trees.

The generalisation error rate from random forests tends to compare favourably with boosting approaches (see Chapter 13), yet the approach tends to be more robust to noise in the training dataset and so tends to be a very stable model builder, as it does not suffer the sensitivity to noise in a dataset that single-decision-tree induction does. The general observation is that the random forest model builder is very competitive with nonlinear classifiers such as artificial neural nets and support vector machines. However, performance is often dataset-dependent, so it remains useful to try a suite of approaches.

Each decision tree is built from a random subset of the training dataset, using what is called replacement (thus it is doing what is known as bagging) in performing this sampling. That is, some observations will be included more than once in the sample, and others won't appear at all. Generally, about two-thirds of the observations will be included in the subset of the training dataset and one-third will be left out.

In building each decision tree model based on a different random subset of the training dataset a random subset of the available variables is used to choose how best to partition the dataset at each node. Each decision tree is built to its maximum size, with no pruning performed. Together, the resulting decision tree models of the forest represent the final ensemble model, where each decision tree votes for the result, and the majority wins. (For a regression model, the result is the average value over the ensemble of regression trees.)

In building the random forest model, we have options to choose the number of trees, the training dataset sample size for building each decision tree, and the number of variables to randomly select when considering how to partition the training dataset at each node. The random forest model builder can also report on the input variables that are actually most important in determining the values of the output variable.

By building each decision tree to its maximal depth (i.e., by not pruning the decision tree), we can end up with a model that is less biased. The randomness introduced by the random forest model builder in selecting the dataset and the variable delivers considerable robustness to noise, outliers, and overfitting when compared with a single tree classifier.

The randomness also delivers substantial computational efficiencies. In building a single decision tree, the model builder may select a random subset of the training dataset. Also, at each node in the process of building the decision tree, only a small fraction of all of the available variables are considered when determining how best to partition the dataset. This substantially reduces the computational requirement.

In summary, a random forest model is a good choice for model building for a number of reasons. First, just like decision trees, very little, if any, preprocessing of the data needs to be performed. The data does not need to be normalised and the approach is resilient to outliers. Second, if we have many input variables, we generally do not need to do any variable selection before we begin model building. The random forest model builder is able to target the most useful variables. Third, because many trees are built and there are two levels of randomness, and each tree is effectively an independent model, the model builder tends not to overfit to the training dataset. A key factor about a random forest being a collection of many decision trees is that each decision tree is not influenced by the other decision trees when constructed.

12.8   Command Summary

This chapter has referenced the following R packages, commands, functions, and datasets:

cforest()	function	Build a conditional random forest.
ctree()	function	Build a conditional inference tree.
evalq()	function	Access environment for storing data.
na.roughfix()	function	Impute missing values.
party	package	Conditional inference trees.
randomForest()	function	Implementation of random forests.
randomForest	package	Build ensemble of decision trees.
str()	function	Show the structure of an object.
weather	dataset	Sample dataset from **rattle**.

Chapter 13

Boosting

The Boosting meta-algorithm is an efficient, simple, and easy to use approach to building models. The popular variant called AdaBoost (an abbreviation for adaptive boosting) has been described as the "best off-the-shelf classifier in the world" (attributed to Leo Breiman by Hastie et al. (2001, p. 302)).

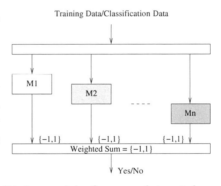

Boosting algorithms build multiple models from a dataset by using some other learning algorithm that need not be a particularly good learner. Boosting associates weights with observations in the dataset and increases (boosts) the weights for those observations that are hard to model accurately. A sequence of models is constructed, and after each model is constructed the weights are modified to give more weight to those observations that are harder to classify. In fact, the weights of such observations generally oscillate up and down from one model to the next. The final model is then an additive model constructed from the sequence of models, each model's output being weighted by some score. There is little tuning required and little is assumed about the learner used, except that it should be a weak learner! We note that boosting can fail to perform if there is insufficient data or if the weak models are overly complex. Boosting is also susceptible to noise.

Boosting algorithms are therefore similar to random forests in that an ensemble of models is built and then combined to deliver a better model

than any of the constituent models. The basic distinguishing characteristic of the boosting approach is that the trees are built one after
another, with refinement being based on the previously built models.
The concept is that after building one model any observations that are
incorrectly classified by that model are boosted. A boosted observation
is essentially given more prominence in the dataset, making the single
observation overrepresented. This has the effect that the next model
is more likely to correctly classify that observation. If not, then that
observation will again be boosted.

In common with random forests, the boosting algorithms tend to be
meta-algorithms. Any type of modelling approach might be used as the
learning algorithm, but the decision tree algorithm is the usual approach.

13.1   Knowledge Representation

The boosting algorithm is commonly presented in terms of decision trees
as their primary form for the representation of knowledge. The key extension to the knowledge representation is in the way that we combine
the decisions that are made by the individual "experts" or models. For
boosting, a weighted score is used, with each of the models in the ensemble having a weight corresponding to the quality of its expertise (e.g.,
the measured accuracy of the individual tree).

13.2   Algorithm

As a meta-learner, boosting employs any simple learning algorithm to
build multiple models. Boosting often relies on the use of a weak learning
algorithm—essentially any weak learner can be used. An ensemble of
weak learners can lead to a strong model.

A weak learning algorithm is one that is only slightly better than
random guessing in terms of error rates (i.e., the model gets the decision
wrong just less than 50% of the time). An early example was a decision
tree of depth 1 (having a single split point and thus often referred to as
a *decision stump*). Each weak model is slightly better than random but
as an ensemble delivers considerable accuracy.

The algorithm begins quite simply by building a "weak" initial model
from the training dataset. Then, any observations in the training data
that the model incorrectly classifies will have their importance within the

algorithm boosted. This is done by assigning all observations a weight—all observations might start, for example, with a weight of 1. Weights are boosted through a formula so that those that are wrongly classified by the model will have a weight greater than 1 for the building of the next decision stump.

A new model is built with these boosted observations. We can think of them as the problematic observations. The algorithm needs to take into account the weights of the observations in building a model. Consequently, the model builder effectively tries harder each iteration to correctly classify these "difficult" observations.

The process of building a model and then boosting observations incorrectly classified is repeated until a newly generated model performs no better than random. The result is then an ensemble of models to be used to make a decision on new data. The decision is arrived at by combining the "expertises" of each model in such a way that the more accurate models carry more weight.

We can illustrate the process abstractly with a simple example. Suppose we have ten observations. Each observation will get an initial weight of, let's say, $\frac{1}{10}$, or 0.1. We build a decision tree that incorrectly classifies four observations (e.g., observations 7, 8, 9, and 10). We can calculate the sum of the weights of the misclassified observations as 0.4 (and generally we denote this as ϵ). This is a measure of the accuracy (actually the inaccuracy) of the model.

The ϵ is transformed into a measure used to update the weights and to provide a weight for the model when it forms part of the ensemble. This transformed value is α and is often something like $0.5 * \log(\frac{1-\epsilon}{\epsilon})$. The new weights for the misclassified observations can then be recalculated as e^{α} times the old weight. In our example, $\alpha = 0.2027$ (i.e., $(0.5 * \log(\frac{1-0.4}{0.4}))$ and so the new weights for observations 7, 8, 9, and 10 become $0.1 * e^{\alpha}$, or 0.1225.

The tree builder is called upon again, noting that some observations are effectively multiplied to have more representation in the algorithm. Thus a different tree is likely to be built that is more likely to correctly classify the observations that have higher weights (i.e., have more representation in the training dataset).

This new model will again have errors. Suppose this time that the model incorrectly classifies observations 1 and 8. Their current weights are 0.1 and 0.1225, respectively. Thus, the new ϵ is $0.1 + 0.1225$, or 0.2225. The new α is then 0.6257. This is the weight that we give to this model when included in the ensemble. It is again used to modify the weights of the incorrectly classified observations, so that observation 1 gets a weight of $0.1 * e^\alpha$, or 0.1869 and observation 8's weight becomes $0.1225 * e^\alpha$, or 0.229. So we can see that observation 8 has the highest weight now since it seems to be quite a problematic observation. The process continues until the individual tree that is built has an error rate of greater than 50%.

To deploy the individual models as an ensemble, each tree is used to classify a new observation. Each tree will provide a probability that it will rain tomorrow (a number between 0 and 1). For each tree, this is multiplied by the weight (α) associated with that tree. The final result is then calculated as the average of these predictions.

Actual implementations of the boosting algorithm use variations to the simple approach we have presented here. Variations are found in the formulas for updating the weights and for weighting the individual models. However, the overall concept remains the same.

13.3   Tutorial Example

Building a Model Using Rattle

The Boost option of the Model tab will build an ensemble of decision trees using the approach of boosting misclassified observations. The individual decision trees are built using **rpart**. The results of building the model are shown in the Textview area. Using the *weather* dataset (loaded by default in Rattle if we click Execute on starting up Rattle) we will see the Textview populated as in Figure 13.1.

The Textview begins with the usual summary of the underlying function call to build the model:

```
Summary of the Ada Boost model:


Call:
ada(RainTomorrow ~ .,
    data = crs$dataset[crs$train, c(crs$input,
    crs$target)], control = rpart.control(maxdepth = 30,,
    minsplit = 20, xval = 10),
    iter = 50)
```

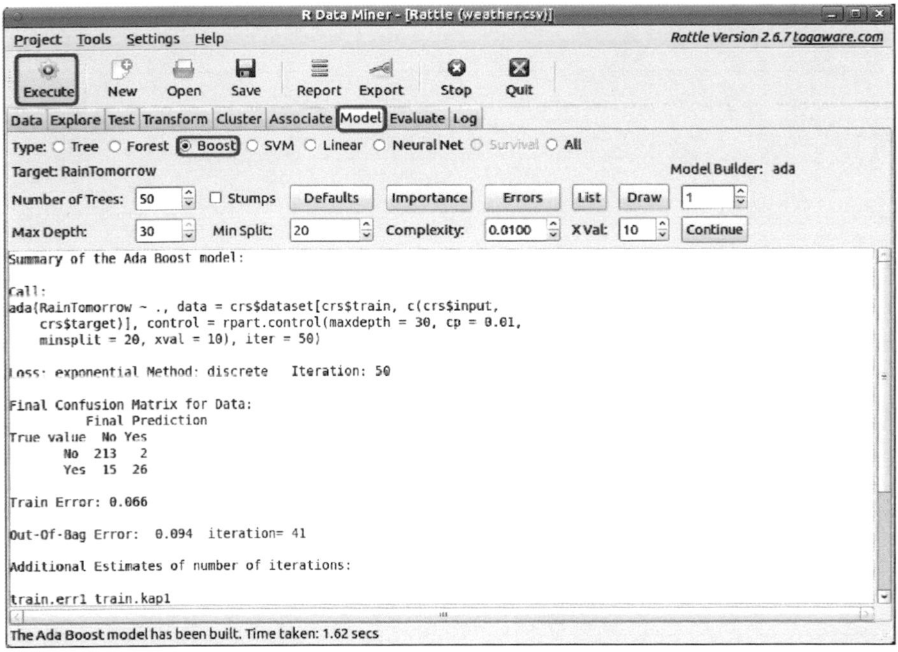

Figure 13.1: Building an AdaBoost predictive model.

The model is to predict RainTomorrow based on the remainder of the variables. The data consists of the dataset loaded into Rattle, retaining only the observations whose index is contained in the training list and including all but columns 1, 2, and 23. The control= argument is passed directly to rpart() and has the same meaning as for rpart() (see Chapter 11). The number of trees to build is specified by the iter= argument. The next line of information reports on some of the parameters used for building the model.

We won't go into the details of the Loss and Method. Briefly, though, the Loss is exponential, indicating that the algorithm is minimising a so called exponential loss function, and the Method used in the algorithm is discrete rather than gentle or real. The Iteration: item simply indicates the number of trees that were asked to be built.

Performance Evaluation

A confusion matrix presents the performance of the model over the training data, and the following line in the Textview reports the training dataset error.

```
Final Confusion Matrix for Data:
          Final Prediction
True value   No Yes
        No  213   2
       Yes   15  26
```

The out-of-bag error and the associated iteration are then reported. This is followed by suggestions of the number of iterations based on the training error and an error measure based on the Kappa statistic. The Kappa statistic adjusts for the situation where there are very different numbers of observations for each value of the target variable. Using these error estimates, the best number of iterations is suggested:

```
Out-Of-Bag Error:  0.094  iteration= 41

Additional Estimates of number of iterations:

train.err1 train.kap1
       47         47
```

The actual training and Kappa (adjusted) error rates are then recorded:

```
train.err train.kap
    0.07      0.28
```

Time Taken

The `ada()` implementation takes longer than `randomForest()` because it is relying on using the inbuilt `rpart()` rather than specially written Fortran code as is the case for `randomForest()`.

```
Time taken: 1.62 secs
```

Error Plot

Once a boosted model has been built, the Error button will display a plot of the decreasing error rate as more trees are added to the model. The plot annotates the curve with a series of five 1s simply to identify the curve. (Extended plots can include curves for test datasets.)

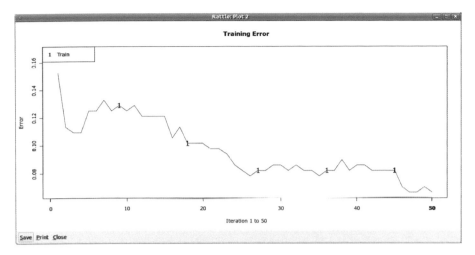

Figure 13.2: The error rate as more trees are added to the ensemble.

Figure 13.2 shows the decreasing error as more trees are added. The plot is typical of ensembles where the error rate drops off quite quickly early on and then flattens out as we proceed. We might decide, from the plot, a point at which we stop building further trees. Perhaps that is around 40 trees for our data.

Variable Importance

A measure of the importance of variables is also provided by **ada** (Culp et al., 2010). Figure 13.3 shows the plot. The measure is a relative measure so that the order and distance between the scores are more relevant than the actual scores.

The measure calculates, for each tree, the improvement in accuracy that the variable chosen to split the dataset offers the model. This is then averaged over all trees in the ensemble.

Of the five most important variables, we notice that there are two categoric variables (`WindDir9am` and `WindDir3pm`). Because of the nature of how variables are chosen for a decision tree algorithm, there may well be a bias here in favour of categoric variables, so we might discount their importance. See Chapter 12, page 265 for a discussion of the issue.

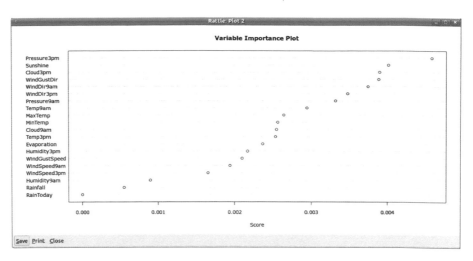

Figure 13.3: The variable importance plot for a boosted model.

Tuning Options

A few basic tuning options for boosting are provided by the Rattle interface. The first option is the Number of Trees to build, which is set to 50 by default. The Max Depth, Min Split, and Complexity are as provided by the decision tree algorithm and are discussed in Section 13.4.

Adding Trees

The Continue button allows further trees to be added to the model. This allows us to easily explore whether the addition of further trees will offer much improvement in the performance of the model, without starting the modelling over again.

To add further trees, increase the value specified in the Number of Trees text box and click the Continue button. This will pick up the model building from where it left off and build as many more trees as is needed to get up to the specified number of trees.

R

The package **ada** provides ada(), which implements the boosting algorithm deployed by Rattle. The ada() function itself uses rpart() from **rpart** to build the decision trees. With the default settings, a very reasonable model can be built.

We will step through the simple process of building a boosted model. First, we create the dataset object, as usual. This will encapsulate the *weather* dataset from **rattle**, together with a collection of other useful data about the *weather* dataset. A training sample is also identified/

```
> library(rattle)
> weatherDS <- new.env()
> evalq({
    data <- weather
    nobs <- nrow(weather)
    vars <- -grep('^(Date|Location|RISK_)', names(data))
    form <- formula(RainTomorrow ~ .)
    target <- all.vars(form)[1]
    set.seed(42)
    train <- sample(nobs, 0.7*nobs)
  }, weatherDS)
```

We can now build the boosted model based on this dataset. Once again we create a container for the model, and include the above container for the dataset within this container.

```
> library(ada)
> weatherADA <- new.env(parent=weatherDS)
```

Within this new container we now build our model.

```
> evalq({
    control <- rpart.control(maxdepth=30,
                             cp=0.010000,
                             minsplit=20,
                             xval=10)
    model <- ada(formula=form,
                 data=data[train, vars],
                 control=control,
                 iter=50)
  }, weatherADA)
```

We can obtain a basic overview of the model simply by printing its value, as we do in the following code block (note that the results here may vary slightly between 32 bit and 64 bit implementations of R).

```
> weatherADA$model

Call:
ada(form, data = data[train, vars],
  control = control, iter = 50)


Loss: exponential Method: discrete    Iteration: 50

Final Confusion Matrix for Data:
         Final Prediction
True value  No Yes
      No   213   2
      Yes   16  25


Train Error: 0.07

Out-Of-Bag Error:  0.105  iteration= 38

Additional Estimates of number of iterations:

train.err1 train.kap1
        41          41
```

The summary() command provides a little more detail.

```
> summary(weatherADA$model)

Call:
ada(form, data = data[train, vars],
  control = control, iter = 50)


Loss: exponential Method: discrete    Iteration: 50

Training Results

Accuracy: 0.93 Kappa: 0.697
```

Replicating AdaBoost Directly using rpart()

We can replicate the boosting process directly using rpart(). We will illustrate this as an example of a little more sophistication in R coding.

We will first load the *weather* dataset and extract the input variables (x) and the output variable (y). To simplify some of the mathematics we will map the predictions to $-1/1$ rather than $0/1$ (since then a model that predicts a value greater than 0 is a positive example and one below zero is a negative example). The data is encapsulated within a container called weatherBRP.

```
> library(rpart)
> weatherBRP <- new.env()
> evalq({
    data <- weather
    vars <- -grep('^(Date|Location|RISK_)', names(data))
    target <- "RainTomorrow"
    N <- nrow(data)
    M <- ncol(data) - length(vars)
    data$Target <- rep(-1, N)
    data$Target[data[target] == "Yes"] <- 1
    vars <- c(vars, -(ncol(data)-1)) # Remove old target
    form <- formula(Target ~ .)
    target <- all.vars(form)[1]
  }, weatherBRP)
```

The first few observations show the mapping from the original target, which has the values No and Yes, to the new numeric values -1 and 1.

```
> head(weatherBRP$data[c("RainTomorrow", "Target")])

  RainTomorrow Target
1          Yes      1
2          Yes      1
3          Yes      1
4          Yes      1
5           No     -1
6           No     -1
```

We can check the list of variables available (only checking the first few here), and note that we exclude four from our analysis:

```
> head(names(weatherBRP$data))

[1] "Date"          "Location"     "MinTemp"
[4] "MaxTemp"       "Rainfall"     "Evaporation"

> weatherBRP$vars

[1]   -1   -2  -23 -24
```

Now we can initialise the observation weights, which to start with are all the same $(1/N)$:

```
> evalq({
    w <- rep(1/N, N)
  }, weatherBRP)
> round(head(weatherBRP$w), 4)

[1] 0.0027 0.0027 0.0027 0.0027 0.0027 0.0027
```

Next we build the first model. The rpart() function, conveniently, has a weights argument, and we simply pass to it the calculated weights store in w. We also set up rpart.control() for building a decision tree stump. The control simply includes maxdepth=, setting it to 1 so that a single-level tree is built:

```
> evalq({
    control <- rpart.control(maxdepth=1)
    M1 <- rpart(formula=form,
                data=data[vars],
                weights=w/mean(w),
                control=control,
                method="class")
  }, weatherBRP)
```

We can then display the first model:

```
> weatherBRP$M1

n= 366

node), split, n, loss, yval, (yprob)
      * denotes terminal node

1) root 366 66 -1 (0.8197 0.1803)
  2) Humidity3pm< 71.5 339 46 -1 (0.8643 0.1357) *
  3) Humidity3pm>=71.5 27  7 1 (0.2593 0.7407) *
```

We see that the decision tree algorithm has chosen Humidity3pm on which to split the data, at a split point of 71.5. For $Humidity3pm < 71.5$ the decision is -1 with probability 0.86, and for $Humidity3pm \geq 71.5$ the decision is 1 with probability 0.75.

We now need to find those observations that are incorrectly classified by the model. The R code here calls predict() to apply the model M1 to the dataset from which it was built. From this result, we get the second column. This is the list of probabilities of each observation being in class 1. If this probability is above 0.5, then the result is 1, otherwise it is -1 (multiplying the logical value by 2 and then subtracting 1 achieves this since TRUE is regarded as 1 and FALSE as 0). The resulting class is then compared with the target, and which() returns the index of those observations for which the prediction differs from the actual class:

```
> evalq({
    ms <- which(((predict(M1)[,2]>0.5)*2)-1 !=
                data[target])
    names(ms) <- NULL
  }, weatherBRP)
```

The indexes of the first few of the 53 misclassified can be listed:

```
> evalq({
    cat(paste(length(ms),
              "observations incorrectly classified:\n"))
    head(ms)
  }, weatherBRP)

53 observations incorrectly classified:
[1]   1  2  3  4  9 17
```

We now calculate the model weight (based on the weighted error rate of this deicison tree) dividing by the total sum of weights to get a normalised value (so that sum(w) remains 1):

```
> evalq({e1 <- sum(w[ms])/sum(w); e1}, weatherBRP)

[1] 0.1448
```

The adjustment is then calculated:

```
> evalq({a1 <- log((1-e1)/e1); a1}, weatherBRP)

[1] 1.776
```

We then update the observation weights:

```
> round(head(weatherBRP$w[weatherBRP$ms]), 4)

[1] 0.0027 0.0027 0.0027 0.0027 0.0027 0.0027
> evalq({w[ms] <- w[ms]*exp(a1)}, weatherBRP)
> round(head(weatherBRP$w[weatherBRP$ms]), 4)

[1] 0.0161 0.0161 0.0161 0.0161 0.0161 0.0161
```

A second model can now be built:

```
> evalq({
    M2 <- rpart(formula=form,
                data=data[vars],
                weights=w/mean(w),
                control=control,
                method="class")
  }, weatherBRP)
```

This results in a simple decision tree involving the variable `Pressure3pm`:

```
> weatherBRP$M2

n= 366

node), split, n, loss, yval, (yprob)
      * denotes terminal node

1) root 366 170.50 -1 (0.5341 0.4659)
  2) Pressure3pm>=1016 206  29.96 -1 (0.8065 0.1935) *
  3) Pressure3pm< 1016 160  70.58 1 (0.3343 0.6657) *
```

Once again we identify the misclassified observations

```
> evalq({
    ms <- which(((predict(M2)[,2]>0.5)*2)-1 !=
                data[target])
    names(ms) <- NULL
  }, weatherBRP)
```

There are 118 of them:

```
> evalq({length(ms)}, weatherBRP)
[1] 118
```

The indexes of the first few can also be listed:

```
> evalq({head(ms)}, weatherBRP)
[1]  9 14 15 16 18 19
```

We again boost the misclassified observations, first calculating the weighted error rate of the decision tree:

```
> evalq({e2 <- sum(w[ms])/sum(w); e2}, weatherBRP)
[1] 0.2747
```

The adjustment is calculated:

```
> evalq({a2 <- log((1-e2)/e2); a2}, weatherBRP)
[1] 0.9709
```

The adjustments are then made to the weights of the individual observations that were misclassified:

```
> round(head(weatherBRP$w[weatherBRP$ms]), 4)

[1] 0.0161 0.0027 0.0027 0.0027 0.0027 0.0027

> evalq({w[ms] <- w[ms]*exp(a2)}, weatherBRP)
> round(head(weatherBRP$w[weatherBRP$ms]), 4)

[1] 0.0426 0.0072 0.0072 0.0072 0.0072 0.0072
```

A third (and for our purposes the last) model can then be built:

```
> evalq({
    M3 <- rpart(formula=form,
                data=data[vars],
                weights=w/mean(w),
                control=control,
                method="class")
    ms <- which(((predict(M3)[,2]>0.5)*2)-1 !=
                data[target])
    names(ms) <- NULL
  }, weatherBRP)
```

Again we identify the misclassified observations:

```
> evalq({length(ms)}, weatherBRP)

[1] 145
```

Calculate the error rate:

```
> evalq({e3 <- sum(w[ms])/sum(w); e3}, weatherBRP)

[1] 0.3341
```

Calculate the adjustment:

```
> evalq({a3 <- log((1-e3)/e3); a3}, weatherBRP)

[1] 0.6896
```

We can then finally adjust the weights (in case we decide to continue building further decision trees):

```
> round(head(weatherBRP$w[weatherBRP$ms]), 4)

[1] 0.0161 0.0161 0.0027 0.0027 0.0027 0.0027

> evalq({w[ms] <- w[ms]*exp(a3)}, weatherBRP)
> round(head(weatherBRP$w[weatherBRP$ms]), 4)

[1] 0.0322 0.0322 0.0054 0.0054 0.0054 0.0054
```

The final (combined or ensemble) model, if we choose to stop here, is then

$$\mathcal{M}(x) = 1.7759 * \mathcal{M}_1(x) + 0.9709 * \mathcal{M}_2(x) + 0.6896 * \mathcal{M}_3(x).$$

13.4 Tuning Parameters

A number of options are given by Rattle for boosting a decision tree model. We briefly review them here.

Number of Trees iter=50

The number of trees to build is specified by the iter= argument. The default is to build 50 trees.

Bagging bag.frac=0.5

Bagging is used to randomly sample the supplied dataset. The default is to select a random sample from the population of 50%.

13.5 Discussion

References

Boosting originated with Freund and Schapire (1995). Building a collection of models into an ensemble can reduce misclassification error, bias, and variance (Bauer and Kohavi, 1999; Schapire et al., 1997). The original formulation of the algorithm adjusts all weights each iteration weights are increased if the corresponding record is misclassified or decreased if it is correctly classified. The weights are then further normalised each iteration to ensure they continue to represent a distribution (so that $\sum_{j=1}^{n} w_j = 1$). This can be simplified, as by Hastie et al. (2001), to increase only the weights of the misclassified observations.

A number of R packages implement boosting. We have covered **ada** here, and this is the package presently used by Rattle. **caTools** (Tuszynski, 2009) provides `LogitBoost()`, which is simple to use and an efficient implementation for very large datasets, using a carefully crafted implementation of decision stumps as the weak learners. **gbm** (Ridgeway, 2010) implements generalised boosted regression, providing a more widely applicable boosting algorithm. **mboost** (Hothorn et al., 2011) is another alternative offering model-based boosting. The variable importance measure implemented for `ada()` is described by Hastie et al. (2001, pp. 331–332).

Alternating Decision Trees—Using Weka

An alternating decision tree (Freund and Mason, 1997), combines the simplicity of a single decision tree with the effectiveness of boosting. The knowledge representation combines tree stumps, a common model deployed in boosting, into a decision tree type structure.

A key characteristic of the tree representation is that the different branches are no longer mutually exclusive. The root node is a prediction node and has just a numeric score. The next layer of nodes are decision nodes and are essentially a collection of decision tree stumps. The next layer then consists of prediction nodes, and so on, alternating between prediction nodes and decision nodes.

A model is deployed by identifying the possibly multiple paths from the root node to the leaves, through the alternating decision tree, that correspond to the values for the variables of an observation to be classified. The observation's classification score (or measure of confidence) is the sum of the prediction values along the corresponding paths.

The alternating decision tree algorithm is implemented in the Weka data mining suite. Weka is available directly from R through **RWeka** (Hornik et al., 2009), which provides its comprehensive collection of data mining tools within the R framework. A simple example will illustrate the incredible power that this offers—using R as a unifying interface to an extensive collection of data mining tools.

We can build an alternating decision tree in R using **RWeka** after installing the appropriate Weka package:

```
> library(RWeka)
> WPM("refresh-cache")
> WPM("install-package", "alternatingDecisionTrees")
```

We use make_Weka_classifier() to turn a Weka object into an R function:

```
> WPM("load-package", "alternatingDecisionTrees")
> cpath <- "weka/classifiers/trees/ADTree"
> ADT <- make_Weka_classifier(cpath)
```

We can obtain some background information about the resulting function by printing the value of the resulting variable:

```
> ADT
An R interface to Weka class
'weka.classifiers.trees.ADTree', which has
information

  Class for generating an alternating decision
  tree. The basic algorithm is based on:

  [...]

Argument list:
  (formula, data, subset, na.action, control =
  Weka_control(),
  options = NULL)

Returns objects inheriting from classes:
  Weka_classifier
```

The function WOW(), standing for "Weka option wizard", will list the command line arguments that become available with the generated function, as seen in the following code block:

```
> WOW(ADT)

-B      Number of boosting iterations.  (Default =
        10)
        Number of arguments: 1.
-E      Expand nodes: -3(all), -2(weight),
        -1(z_pure), >=0 seed for random walk
        (Default = -3)
        Number of arguments: 1.
-D      Save the instance data with the model
```

Next we perform our usual model building. As always we first create a container for the model, making available the appropriate dataset container for use from within this new container:

```
> weatherADT <- new.env(parent=weatherDS)
```

The model is built as a simple call to ADT:

```
> evalq({
    model <- ADT(formula=form, data=data[train, vars])
  }, weatherADT)
```

The resulting alternating decision tree can then be displayed as we see in the following code block.

```
> weatherADT$model

Alternating decision tree:

: -0.794
|   (1)Pressure3pm < 1011.9: 0.743
|   (1)Pressure3pm >= 1011.9: -0.463
|   |   (3)Temp3pm < 14.75: -1.498
|   |   (3)Temp3pm >= 14.75: 0.165
|   (2)Sunshine < 8.85: 0.405
|   |   (4)WindSpeed9am < 6.5: 0.656
|   |   (4)WindSpeed9am >= 6.5: -0.26
|   |   |   (8)Sunshine < 6.55: 0.298
|   |   |   |   (9)Temp3pm < 18.75: -0.595
|   |   |   |   (9)Temp3pm >= 18.75: 0.771
|   |   |   (8)Sunshine >= 6.55: -0.931
|   (2)Sunshine >= 8.85: -0.76
|   |   (6)MaxTemp < 24.35: -1.214
|   |   (6)MaxTemp >= 24.35: 0.095
|   |   |   (7)Sunshine < 10.9: 0.663
|   |   |   (7)Sunshine >= 10.9: -0.723
|   (5)Pressure3pm < 1016.1: 0.295
|   (5)Pressure3pm >= 1016.1: -0.658
|   |   (10)MaxTemp < 19.55: 0.332
|   |   (10)MaxTemp >= 19.55: -1.099
Legend: -ve = No, +ve = Yes
Tree size (total number of nodes): 31
Leaves (number of predictor nodes): 21
```

We can then explore how well the model performs:

```
> evalq({
    predictions <- predict(model, data[-train, vars])
    table(predictions, data[-train, target],
        dnn=c("Predicted", "Actual"))
  }, weatherADT)

          Actual
Predicted No Yes
      No  72  11
      Yes 13  14
```

Compare this with the `ada()` generated model:

```
> evalq({
    predictions <- predict(model, data[-train, vars])
    table(predictions, data[-train, target],
        dnn=c("Predicted", "Actual"))
  }, weatherADA)

          Actual
Predicted No Yes
      No  78  10
      Yes  7  15
```

In this example, the `ada()` model performs better than the `ADT()` model.

13.6   Summary

Boosting is an efficient, simple, and easy-to-understand model building strategy that tends not to overfit our data, hence building good models. The popular variant called AdaBoost (an abbreviation for adaptive boosting) has been described as the "best off-the-shelf classifier in the world" (attributed to Leo Breiman by Hastie et al. (2001, p. 302)).

Boosting algorithms build multiple models from a dataset, using some other model builders, such as a decision tree builder or neural network, that need not be particularly good model builders. The basic idea of boosting is to associate a weight with each observation in the dataset. A series of models are built and the weights are increased (boosted) if a model incorrectly classifies the observation. The weights of such observations generally oscillate up and down from one model to the next. The final model is then an additive model constructed from the sequence of models, each model's output weighted by some score. There is little tuning required and little is assumed about the model builder used, except that it should be relatively weak model. We note that boosting can fail to perform if there is insufficient data or if the weak models are overly complex. Boosting is also susceptible to noise.

13.7   Command Summary

This chapter has referenced the following R packages, commands, functions, and datasets:

ada()	function	Implementation of AdaBoost.
ada	package	Builds AdaBoost models.
caTools	package	Provides LogitBoost().
gbm	package	Generalised boosted regression.
LogitBoost()	function	Alternative boosting algorithm.
predict()	function	Applies model to new dataset.
randomForest()	function	Implementation of random forests.
rattle	package	The *weather* dataset and GUI.
rpart()	function	Builds a decision tree model.
rpart.control()	function	Controls ada() passes to rpart().
rpart	package	Builds decision tree models.
RWeka	package	Interface Weka software.
summary()	function	Summarise an ada model.
which()	function	Elements of a vector that are TRUE.
WOW()	function	The Weka option wizard.

Chapter 14

Support Vector Machines

A support vector machine (SVM) searches for so-called *support vectors* which are observations that are found to lie at the edge of an area in space which presents a boundary between one of these classes of observations (e.g., the squares) and another class of observations (e.g., the circles). In the terminology of SVM we talk about the space between these two

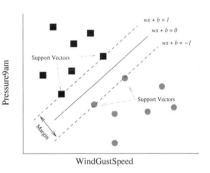

regions as the *margin* between the classes. Each region contains observations with the same value for the target variable (i.e., the class). The support vectors, and only the support vectors, are used to identify a hyperplane (a straight line in two dimensions) that separates the classes. The maximum margin between the separable classes is sought. This then represents the model.

It is usually quite rare that we can separate the data with a straight line (or a hyperplane when we have more than two input variables). That is, the data is not usually distributed in such a way that it is linearly separable. When this is the case, a technique is used to combine (or remap) the data in different ways, creating new variables so that the classes are then more likely to become linearly separable by a hyperplane (i.e., so that with the new dimensional data there is a gap between observations in the two classes). We can use the model we have built to score new observations by mapping the data in the same way as when the model was built, and then decide on which side of the hyperplane the observation lies and hence the decision associated with it.

Support vector machines have been found to perform well on problems that are nonlinear, sparse, and high-dimensional. A disadvantage is that the algorithm is sensitive to the choice of tuning option (e.g., the type of transformations to perform), making it harder to use and time-consuming to identify the best model. Another disadvantage is that the transformations performed can be computationally expensive and are performed both whilst building the model and when scoring new data.

An advantage of the method is that the modelling only deals with these support vectors rather than the whole training dataset, and so the size of the training set is not usually an issue. Also, as a consequence of only using the support vectors to build a model, the model is less affected by outliers.

14.1   Knowledge Representation

The approach taken by a support vector machine model is to build a linear model; that is, to identify a line (in two dimensions) or a flat plane (in multiple dimensions) that separates observations with different values of the target variable. If we can find such a line or plane, then we find one that maximises the area between the two groups (when we are looking at binary classification, as with the *weather* dataset).

Consider a simple case where we have just two input variables (i.e., two-dimensional space). We will choose `Pressure3pm` and `Sunshine`. We will also purposefully select observations that will clearly demonstrate a significant separation (margin) between the observations for which it rains tomorrow and those for which it does not. The R code here illustrates our selection of the data and drawing of the plot. From the *weather* dataset, we only select observations that meet a couple of conditions to get two clumps of observations, one with `No` and the other with `Yes` for `RainTomorrow` (see Figure 14.1):

```
> library(rattle)
> obs <- with(weather, Pressure3pm+Sunshine > 1032 |
              (Pressure3pm+Sunshine < 1020 &
              RainTomorrow == "Yes"))
> ds <- weather[obs,]
> with(ds, plot(Pressure3pm, Sunshine,
               pch=as.integer(RainTomorrow),
               col=as.integer(RainTomorrow)+1))
> lines(c(1016.2, 1019.6), c(0, 12.7))
> lines(c(1032.8, 1001.5), c(0, 12.7))
> legend("topleft", c("Yes", "No"), pch=2:1, col=3:2)
```

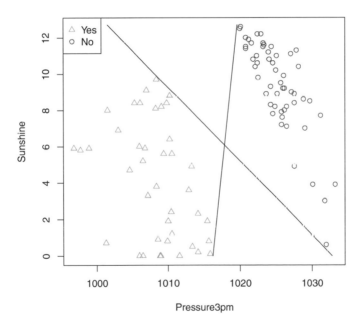

Figure 14.1: A simple and easily linearly separable collection of observations.

Two lines are shown in Figure 14.1 as two possible linear models. Taking either line as a model, the observations to the left will be classified as Yes and those to the right as No. However, there is an infinite collection of possible lines that we could draw to separate the two regions.

The support vector approach suggests that we find a line in the separating region such that we can make the line as thick as possible to butt
up against the observations on the boundary. We choose the line that
fills up the maximum amount of space between the two regions, as in
Figure 14.2. The observations that butt up against this region are the
support vectors.

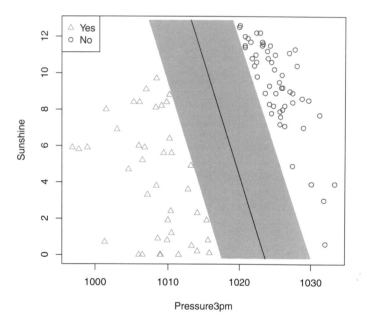

Figure 14.2: A maximal region or margin between the two classes of observations.

This is the representation of the model that we build using the approach of identifying support vectors and a maximal region between the
classifications. The approach generalises to multiple dimensions (i.e.,
many input variables), where we search for hyperplanes that maximally
fill the space between classes in the same way.

14.2   Algorithm

It is rarely the case that our observations are linearly separable. More likely, the data will appear as it does in Figure 14.3, which was generated directly from the data.

```
> ds <- weather
> with(ds, plot(Pressure3pm, Sunshine,
                pch=as.integer(RainTomorrow),
                col=as.integer(RainTomorrow)+1))
> legend("topleft", c("Yes", "No"), pch=2:1, col=3:2)
```

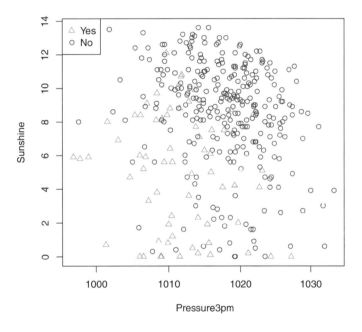

Figure 14.3: A nonlinearly separable collection of observations.

This kind of situation is where the kernel functions play a role. The idea is to introduce some other derived variables that are obtained from the input variables but combined and mathematically changed in some nonlinear way. Rather simple examples could be to add a new variable which squares the value of `Pressure3pm` and another new variable that multiplies `Pressure3pm` by `Sunshine`. Adding such variables to the data can enhance separation. Figure 14.4 illustrates the resulting location of the observations, showing an improvement in separation (though to artificially exaggerate the improvement, not all points are shown).

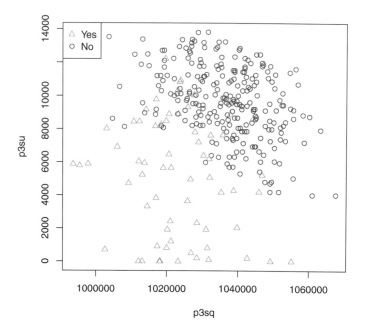

Figure 14.4: Nonlinearly transformed observations showing `Pressure3pm` squared (x-axis) against `Pressure3pm` multiplied by `Sunshine`, artificially enhanced.

A genuine benefit is likely to be seen when we add further variables to our dataset. It becomes difficult to display such multi-dimensional plots on the page, but tools like GGobican assist in visualising such data and confirming improved separation.

The basic kernel functions that are often used are the radial basis function, a linear function, and a polynomial function. The radial basis function uses a formula somewhat of the form $e^{-\gamma\|x-x'\|^2}$ for two observations x and x' (without going into the details). The polynomial function has the form $(1+\langle x, x'\rangle)^d$, for some integer d. For two input variables X_1

and X_2 and a power of 2, this becomes $(1+x_1x_1'+x_2x_2')^2$. Again, we skip the actual details of how such a formula is used. There are a variety of kernel functions available, but the commonly preferred one, and a good place to start, is the radial basis function.

Once the input variable space has been appropriately transformed, we then proceed to build a linear model as described in Section 14.1.

14.3   Tutorial Example

Building a Model Using Rattle

Rattle supports the building of support vector machine (SVM) models through **kernlab** (Karatzoglou et al., 2004). This package provides an extensive collection of kernel functions that can be applied to the data. This works by introducing new variables. Quite a variety of tuning options are provided by **kernlab**, but only a few are given through Rattle.

It is quite easy to experiment with different kernels using the *weather* dataset provided. The default kernel (radial basis function) is a good starting point. Such models can be quite accurate with no or little tuning. Two parameters are available for the radial basis function. C= is a penalty or cost parameter with a default value of 1. The Options widget can be used to set different values (e.g., C=10).

We review here the information provided to summarise the model, as displayed in Figure 14.5. The Textview begins with the summary of the model, identifying it as an object of class (or type) ksvm (kernel support vector machine):

```
Summary of the SVM model (built using ksvm):

Support Vector Machine object of class "ksvm"
```

The type of support vector machine is then identified. The C-svc indicates that the standard regularised support vector classification (svc) algorithm is used, with parameter C for tuning the algorithm. The value used for C is also reported:

```
SV type: C-svc  (classification)
 parameter : cost C = 1
```

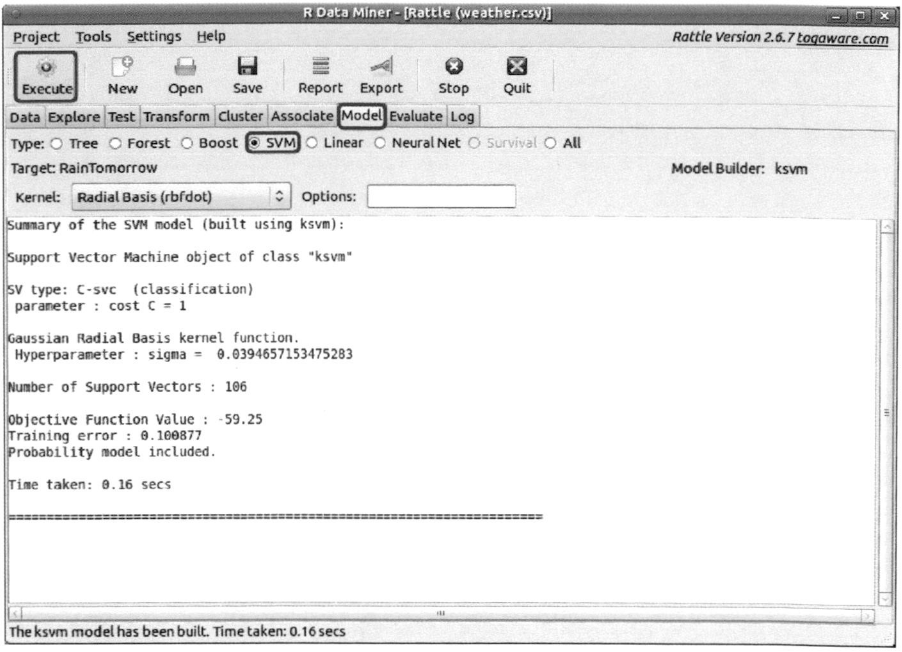

Figure 14.5: Building a support vector machine classification model.

An automatic algorithm is used to estimate another parameter (sigma) for the radial basis function kernel. The next two lines include an indication of the estimated value:

```
Gaussian Radial Basis kernel function.
 Hyperparameter : sigma =  0.0394657153475283
```

The remaining lines report on the characteristics of the model, including how many observations are on the boundary (i.e., the support vectors), the value of the so-called objective function that the algorithm optimises, and the error calculated on the training dataset:

```
Number of Support Vectors : 106

Objective Function Value : -59.25
Training error : 0.100877
Probability model included.
```

Time Taken

The support vector machine is reasonably efficient:

```
Time taken: 0.16 secs
```

R

There is a wealth of functionality provided through **kernlab** and ksvm() for building support vector machine models. We will cover the basic functionality here. As usual, we begin with the dataset object from which we will be building our model:

```
> library(rattle)
> weatherDS <- new.env()
> evalq({
    data <- weather
    nobs <- nrow(weather)
    target <- "RainTomorrow"
    vars <- -grep('^(Date|Location|RISK_)', names(data))
    set.seed(42)
    train <- sample(nobs, 0.7*nobs)
    form <- formula(RainTomorrow ~ .)
  }, weatherDS)
```

We can now build the boosted model based on this dataset:

```
> library(kernlab)
> weatherSVM <- new.env(parent=weatherDS)
> evalq({
    model <- ksvm(form,
                  data=data[train, vars],
                  kernel="rbfdot",
                  prob.model=TRUE)
  }, weatherSVM)
```

The kernel= argument indicates that we will use the radial basis function as the kernel function. The prob.model= argument, set to TRUE, results in a model that predicts the probability of the outcomes. We obtain the usual overview of the model by simply printing its value:

```
> weatherSVM$model

Support Vector Machine object of class "ksvm"

SV type: C-svc  (classification)
 parameter : cost C = 1

Gaussian Radial Basis kernel function.
 Hyperparameter : sigma =  0.0394335857291656

Number of Support Vectors : 107

Objective Function Value : -59.23
Training error : 0.100877
Probability model included.
```

14.4   Tuning Parameters

We describe here a number of tuning parameters, but many other options are available and are documented as part of **kernlab**.

Model Type type=

The ksvm() function can be used for a variety of modelling tasks, depending on the type of target variable. We are generally interested in classification tasks using the so-called C-svc formulation (support vector classification with a C parameter for tuning). This is a standard formulation for SVMs and is referred to as regularised support vector classification. Other options here include nu-svc for automatically regularised support vector classification, one-svc for novelty detection, eps-svr for support vector regression that is robust to small (i.e., epsilon) errors, and nu-svr for support vector regression that automatically minimises epsilon. Other options are available.

Kernel Function kernel=

The kernel method is the mechanism used to map our observations into a higher dimensional space. It is then within this higher dimensional space that the algorithm looks for a hyperplane that partitions our observations to find the maximal margin between the different values of the target variable.

The ksvm() function supports a number of kernel functions. A good starting point is the radial basis function (using a Gaussian type of function). This is the rfdot option. The "dot" refers to the mathematical dot function or inner product between two vectors. This is integral to how support vector machines work, though not covered here. Other options include polydot for a polynomial kernel, vanilladot for a linear kernel, and splinedot for a spline kernel, amongst others.

Class Probabilities prob.model=

If this is set to TRUE, then the resulting model will calculate class probabilities.

Kernel Parameter: Cost of Constraints Violation C=

The cost parameter C= is by default 1. Larger values (e.g., 100 or 10,000) will consider more the points near the decision boundary, whilst smaller values relate to points further away from the decision boundary. Depending on the data, the choice of the cost argument may only play a small role in the resulting model.

Kernel Parameter: Sigma sigma=

For a radial basis function kernel, the sigma value can be set. Rattle uses automatic sigma estimation (using sigest()) for this kernel. This will find a good sigma value based on the data.

To experiment with various sigma values we can use the R code from Rattle's Log tab and paste that into the R Console and then add in the additional settings and run the model. This parameter tunes the kernel function selected, and so is listed as the kparm= list.

Cross Validation cross=

We can specify an integer value here to indicate whether to perform k-fold cross-validation.

14.5 Command Summary

This chapter has referenced the following R packages, commands, functions, and datasets:

kernlab	package	Kernel-based algorithms for machine learning.
ksvm()	function	Build an SVM model.
rattle	package	The *weather* dataset and GUI.
sigest()	function	Sigma estimation for kernel.
weather	dataset	Sample dataset from **rattle**.

Part III

Delivering Performance

Chapter 15

Model Performance Evaluation

If a model looks too good to be true, then generally it is.

The preceding chapters presented a number of algorithms for building descriptive and predictive models. Before we can identify the best from amongst the different models, we must evaluate the performance of the model. This will allow us to understand what to expect when we use the model to score new observations. It can also help identify whether we have made any mistakes in our choice of input variables. A common error is to include as an input variable a variable that directly relates to the outcome (like the amount of rain tomorrow when we are predicting whether it will rain tomorrow). Consequently, this input variable is exceptionally good at predicting the target.

In this chapter, we consider the issue of evaluating the performance of the models that we have built. Essentially, we consider `predict()`, provided by R and accessed through Rattle's Evaluate tab, and the functions that summarise and analyse the results of the predictions.

We will work through each of the approaches for evaluating model performance. We start with a simple table, called a confusion matrix, that compares predictions with actual answers. This also introduces the concepts of true positives, false positives, true negatives, and false negatives. We then explain a risk chart which graphically compares the performance of the model against known outcomes and is used to identify a suitable trade-off between effort and outcomes. Traditional ROC curves are then introduced. We finish with a discussion of simply scoring datasets and saving the results to a file.

In applying a model to a new dataset, the new dataset must contain all of the same variables and have the same data types on which the model was built. This is true even if any variables were not used in the final model. If the variable is missing from the new dataset, then generally an error is generated.

15.1   The Evaluate Tab: Evaluation Datasets

Rattle's Evaluate tab provides access to a variety of options for evaluating the performance of our models. We can see the options listed in Figure 15.1. We briefly introduce the options here and expand on them in this chapter.

Figure 15.1: The Evaluate tab options.

Types of Evaluations

The range of different Types of evaluations is presented as a series of radio buttons running from Confusion Matrix to Score. Only one type of evaluation is permitted to be chosen at any time. Each type of evaluation is presented in the following sections of this chapter.

Models to Evaluate

Below the row of evaluation types is a row of check boxes to choose the models we wish to evaluate. These check boxes are only available once a model has been built. As models are built, the check boxes will become available as options to check.

As we move from the **Model** tab to this **Evaluate** tab, the most recently built model will be automatically checked (and any previously checked model choices will be unselected). This corresponds to a common pattern of behaviour in that often we will build and tune a model, then want to explore its performance by moving to this **Evaluate** tab. If the **All** option has been chosen within the **Model** tab, then all models that were successfully built will automatically be checked on the **Evaluate** tab. This is the case here, where the six predictive models are checked.

Dataset Used for Evaluation

To evaluate a model, we need to identify a dataset on which to perform the evaluation. The next row of options within the **Rattle** interface provides a collection of alternative sources of data.

The first four options for the **Data** correspond to the partitioning of the dataset specified on the **Data** tab. The options are **Training, Validation, Testing,** and **Full**. The concept of a training/validation/testing dataset partition was discussed in Section 3.1, and we discussed the concept of sampling and associated biases in Section 4.7. We now discuss it further in the context of evaluating the models.

The first option (but not the best option) is to evaluate our model on the **Training** dataset. This is generally not a good idea, and an information dialogue will be shown to reinforce this.

The problem with evaluating our model on the training dataset is that we have built it on this training dataset. It is often the case that the model will perform very well on that dataset. It should, because we've tried hard to make sure it does. But this does not give us a very good idea of how well the model will perform in general on previously unseen data.

We need a better guide to how well the model will perform in general, that is, how the model will perform on new and previously unseen data. To answer that question, we need to apply the model to such data. In doing so, we will obtain the overall error rate of the model. This is simply the proportion of observations where the model and the actual known outcomes differ. This error rate, and not the error rate from the training dataset, will then be a better estimate of how well the model will perform. It is a less biased estimate of the error.

We use the **Validation** dataset to test the performance of a model whilst we are building and fine-tuning it. Thus, after building one decision tree, we will check its performance against this validation dataset. We might then change some of the tuning options for building a decision tree model. We compare the new model against the old one based on its performance on the validation dataset. In this sense, the validation dataset is used during the modelling process to build the final model. Consequently, we will still have a biased estimate of the final performance of our model if we rely on the validation dataset for this measure.

The Testing dataset is then a "hold-out" dataset that has not been used at all during the model building. Once we have identified our "best" model based on using the validation dataset, the model's performance on the testing dataset is then assessed. This is then an estimate of the expected performance on any new data.

The fourth option uses the Full dataset for evaluating the model (the combined training, validation, and testing datasets). This might be seen to be useful only as a curiosity rather than for accurate performance.

Another option available as a data source is provided through the Enter choice. This is available when Score is chosen as the type of evaluation. In this case, a window will pop up to allow us to directly enter some data and have that "scored" by the model.

The final two options for the data source are a CSV File and an R Dataset. These allow data to be loaded into R from a CSV file, and the model can be evaluated on that dataset. Alternatively, for a data frame already available through R, the R Dataset will allow it to be chosen and the model evaluated on that.

Risk Variable

The final row of options begins with an informative label that reports on the name of the Risk Variable chosen in the Data tab. The risk variable is used as a measure of how significant each observation is with respect to the target variable. For example, it might record the dollar value of the fraud or the amount of rain received "tomorrow." The risk chart makes use of this variable, if there is one, and it is included in the common area of the Evaluate tab for information purposes only.

Scoring

The remaining options on the final row of options relate to scoring. Many models can predict an outcome or a probability for a particular outcome.

The **Report** option allows us to choose which we would like to see in the output when scoring. The **Include** option indicates whether to include all variables for each observation in the output or just the identifiers (those variables marked as having an Ident role on the **Data** tab).

A Note on Cross-Validation

In Section 2.7, we introduced the concept of partitioning our dataset into three samples: the training, validation, and testing datasets. This concept was then further discussed in Section 3.1 and in the section above. In considering each of the modelling algorithms, we also touched on the evaluation of the models, using the validation dataset, as part of the model building process. We have stressed that the testing dataset is used as the final *unbiased* estimate of the performance of a model.

A related paradigm for evaluating the performance of our models is through the use of *cross-validation*. Indeed, some of the algorithms implemented in R will perform cross-validation for us and report a performance measure based on it. The decision tree algorithm using `rpart()` is an example.

Cross-validation is a simple concept. Given a dataset, we partition it into, perhaps, ten random sample subsets. Then we build a model using nine of those subsets, combined to form the training dataset. We can then measure the performance of the resulting model on the *hold-out dataset*. Then we can repeat this by selecting a different nine subsets to use as a training dataset. Once again, the remaining dataset will serve as a testing dataset. This can be repeated ten times to give us a measure of the expected performance of the resulting model.

A related concept, and one that we often find in the context of ensemble models, is the concept of *out-of-bag*, or *OOB*, measures of performance. We saw this concept when building a random forest model in Section 12.4. We might recall that in building a random forest we sample a subset of the full dataset. The subset is used as the training dataset. Thus, the remaining dataset can be used to test the performance of the resulting model.

In those cases where the R implementation of an algorithm provides its own performance measure, using cross-validation or out-of-bag estimates, we might choose not to create a validation dataset in **Rattle**. Instead, we can rely on the measure supplied by the algorithm as we build and fine-tune our models. The testing dataset remains useful then to provide an unbiased measure once we have built our best models.

15.2   Measure of Performance

Quite a collection of measures has been developed over many years to gauge the performance of a model. The help page for `performance()` of **ROCR** (Sing et al., 2009) in R collects most of them together with a brief description, with 30 other measures listed. To review that list, using the R Console, simply ask for help:

```
> library(ROCR)
> help(performance)
```

We will discuss performance in the context of a binary classification model. This has been our focus with the *weather* dataset, predicting No or Yes for the variable RainTomorrow. For binary classification, we also often identify the predictions as positives or negatives. Thus, in terms of predicting whether it will rain tomorrow, Yes is the *positive* class and No is the *negative* class.

For an evaluation of a model, we apply the model to a dataset of observations with known actual outcomes (classes). The model will be used to predict the class for each observation. We then compare the predicted class with the actual class.

Error Rate

The simplest measure of the performance of a model is the error rate. This is calculated as the proportion of observations for which the model incorrectly predicts the class with respect to the actual class. That is, we simply divide the number of misclassifications by the total number of observations.

True and False Positives and Negatives

If our weather model predicts Yes in agreement with the actual outcome, then we refer to this as a **true positive**. Similarly, when they both agree on the negative, we refer to it as a **true negative**. On the other hand, when the model predicts No and the actual is Yes, then we have a **false negative**. Predicting a Yes when it is actually a No results in a **false positive**.

Often it is useful to differentiate in this way between the types of misclassification errors that a model makes. For example, in the context of our weather dataset, it makes a difference whether we have a false positive or a false negative. A false positive would predict that it will rain tomorrow when in fact it does not. The consequence is that I might take my umbrella with me but I won't need to use it—only a minor inconvenience.

However, a false negative predicts that it won't rain tomorrow but in fact it does rain. Relying on the model, I would not bother with an umbrella. Consequently, I am caught in the rain and get uncomfortably wet. The consequences of a false negative in this case are more significant for me than they are for a false positive.

Whether false positives or false negatives are more of an issue depends on the application. For medical applications, a false positive (e.g., falsely diagnosed with cancer) may be less of an issue than a false negative (e.g., the diagnosis of cancer being missed). Different model builders can deal with these situations in different ways. The decision tree algorithm, for example, can accept a loss matrix that gives different weights to the different outcomes. This will then bias the model building to avoid one type of error or the other.

Often, we are interested in the ratio of the number of true positives to the number of predicted positives. This is referred to as the true positive rate and similarly for the false positive rate and so on.

Precision, Recall, Sensitivity, Specificity

The **precision** of a model is the ratio of the number of true positives to the total number of predicted positives (the sum of the true positives and the false positives). It is a measure of how accurate the positive predictions are, or how *precise* the model is in predicting.

The **recall** of a model is just another name for the true positive rate. It is a measure of how many of the actual positives the model can identify, or how much the model can *recall*. The recall is also known as the **sensitivity** of the model.

Another measure that often arises in the context of sensitivity is **specificity**. This is simply another name for the true negative rate.

Other Measures

We will use and refine the measure we have introduced here in describing the various approaches to evaluating our models in the following sections. As the `help()` for **ROCR** indicates, we have very many to choose from, and which works best for the many different application areas is often determined through trial and error and experience.

15.3 Confusion Matrix

A confusion matrix (also known as an error matrix) is appropriate when predicting a categoric target (e.g., in binary classification models). We saw a number of confusion matrices in Chapter 2.

In Rattle, the Confusion Matrix is the default on the Evaluate tab. Clicking the Execute button will run the selected model(s) against the chosen dataset to predict the outcomes for each of the observations in that dataset. The predictions are compared with the actual observations, and the true and false positives and negatives are calculated.

Figure 15.2 illustrates this for the decision tree model using the *weather* dataset. We see in Figure 15.2 that six models have been built, and the Textview will show the confusion matrix for each of the selected models. A quick way to build each type of model is to choose the All option on the Model tab.

The confusion matrix displays the predicted versus the actual results in a table. The first table shows the actual counts, whilst the second table shows the percentages. For the decision tree applied to the validation dataset, there are 5 true positives and 39 true negatives, and so the model is correct for 44 observations out of 54. That is, the overall error rate is 10 out of 54, or 19%.

The false positives and false negatives have the same count. On five days we will get wet and on another five we will carry an umbrella with us unnecessarily.

If we scroll the text view window of the Evaluate tab, we can see the confusion-matrix-based performance measure for other models. The random forest appears to provide a slightly more accurate prediction, as we see in Figure 15.3.[1]

---

[1] Note that the results vary slightly between different deployments of R, particularly between 64 bit R, as here, and 32 bit R.

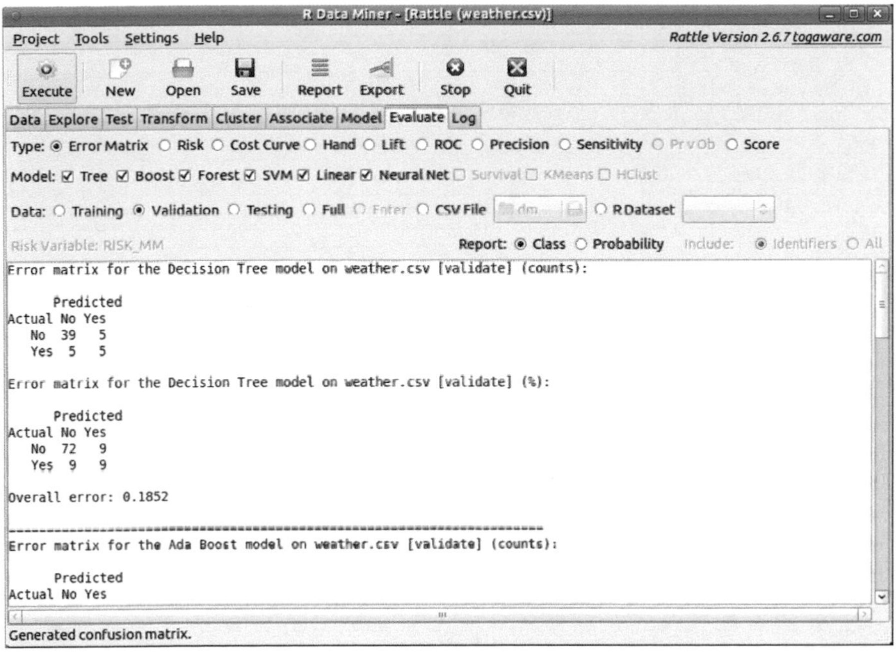

Figure 15.2: The Evaluate tab showing a confusion matrix.

The overall error rate for the random forest is 12%, with 4 true positives and 40 true negatives. Compared with the decision tree, there is one less day when we will get wet and three fewer days when we would unnecessarily carry our umbrella. We might instead look for a model that reduces the false negatives rather than the false positives. (Also remember that we should be careful when comparing such small numbers of observations—the differences won't be significant, though when using very large training datasets, as would be typical for data mining, we are in a better position to compare.)

15.4 Risk Charts

A *risk chart*, also known as a *cumulative gain chart*, provides another perspective on the performance of a binary classification model. Such a chart can be displayed by choosing the Risk option on the Evaluate tab. We will explain risk charts here using the *audit* dataset. The use of risk charts to evaluate models of fraud and noncompliance is more logical

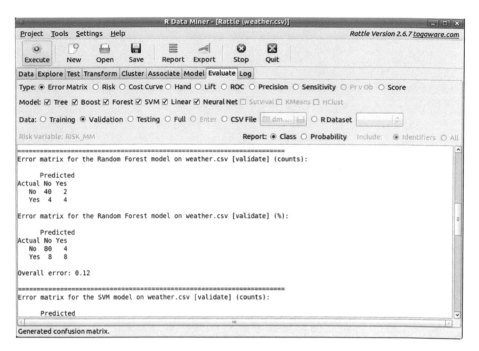

Figure 15.3: Confusion matrix for the random forest model.

than with the application to predicting rain.

The *audit* dataset (Section B.2) contains observations of taxpayers who have been audited together with the outcome of the audit: No or Yes. A positive outcome indicates that the taxpayer was required to update the tax return because of inaccuracies in the figures reported. A negative outcome indicates that the tax return required no adjustment. For each adjustment we also record its dollar amount (as the risk variable).

We can build a random forest model using this dataset, but we first need to load it into **Rattle**. To do so, go back to the **Data** tab and after loading **rattle**'s *weather* dataset click on the **Filename** chooser. We can then select the file **audit.csv**. Click on **Execute** to have the new dataset loaded. Then, from **Rattle**'s **Model** tab, build a **Forest** and then request a **Risk Chart** from the **Evaluate** tab. The resulting risk chart is shown in Figure 15.4. To read the risk chart, we will pick a particular point and consider a specific scenario. The scenario is that of auditing taxpayers. Suppose we normally audit 100,000 taxpayers each year. Of those, only 24,000, let's say, end up requiring an adjustment to their tax return. We call this the *strike rate*. That is, we strike 24,000 out of the 100,000 as being of interest—a strike rate of 24%.

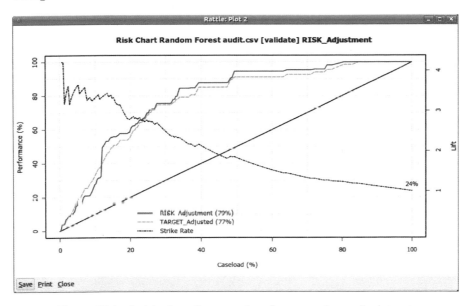

Figure 15.4: A risk chart for a random forest on the *audit* dataset.

Suppose our funding now allows us to audit only 50,000 taxpayers. If we were to randomly select 50% from the 100,000 taxpayers, then we would expect to identify just 50% of the actual taxpayers whose tax returns required an adjustment. That is, we would identify only 12,000 of the 24,000 tax returns requiring an adjustment from amongst the 50,000 taxpayers randomly selected. This random selection is represented by the diagonal line in the plot. A random 50% caseload (i.e., 50,000 cases) will deliver a 50% performance (i.e., only half of the known cases of interest will be found). We can think of this as the baseline—this is what the situation would be if we used random selection and no other model.

We now introduce our random forest model, which predicts the likelihood of a taxpayer's tax return requiring an adjustment. For each taxpayer, the model provides a score—the probability of the taxpayer's tax return requiring an adjustment. We can now prioritise our audits of taxpayers based on these scores so that taxpayers with a higher score are audited before taxpayers with a lower score. Once again, but now using this priority, we choose to audit only 50,000 taxpayers, but we select the 50,000 that have the highest risk scores.

The dashed green line of the plot indicates the performance achieved when using the model to prioritise the audits. For a 50% caseload, the performance is approximately 90%. That is, we expect to identify 90% of the tax returns requiring an adjustment. So 21,600 of the 24,000 known adjustments, from amongst the 50,000 taxpayers chosen, are expected to be identified. That is a significant improvement over the 12,000 from the 50,000 selected randomly. Indeed, as the blue line in the plot indicates, that provides a *lift* in performance of almost 2. That is, we are identifying almost twice as many tax returns requiring adjustment than we would expect if we were simply selecting taxpayers randomly.

In this light, the model provides quite a significant benefit. Note that we are not particularly concentrating on error rates as such but on the benefit we achieve in using the model to rank or prioritise our business processes. Whilst a lot of attention is often paid to simplistic measures of model performance, other factors usually come into play in deciding which model performs best.

Note also from the plot in Figure 15.4 that after we have audited about 85% of the cases (i.e., at a caseload of 85) the model achieves 100% performance. That is, the model has ensured that all tax returns requiring adjustment have been identified by the time we have audited 85,000 taxpayers. A conservative use of the model would then ensure nearly all required audits (i.e., 24,000) are performed, yet saving 15% of the effort previously required to identify all of the required audits. We also note that out of the 85,000 audits we are still unnecessarily auditing 61,000 taxpayers.

The solid red line of the risk chart often follows a path similar to that of the green line. It provides an indication of the measure of the size of the risk covered by the model. It is based on the variable identified as having a role as a **Risk** variable on the **Data** tab. In our case, it is the variable RISK_Adjustment and records the dollar amount of any adjustment made to a tax return. In that sense, it is a measure of the size of the risk.

The "risk" performance line is included for information. It has not been used in the modelling at all (though it could have been). Empirically, we often note that it sits near or above the "target" performance line. If it sits high above the target line, then the model is fortuitously identifying higher-risk cases earlier in the process, which is a useful outcome.

A risk chart can be displayed for any binary classification model built using Rattle. In comparing risk charts for different models, we are looking for a larger area under the curve. This generally means that the curve "closer" to the top left of the risk chart identifies a better-performing model than a curve that is closer to the baseline (diagonal line). Figure 15.5 illustrates the output when multiple models are selected, so that performances can be directly compared.

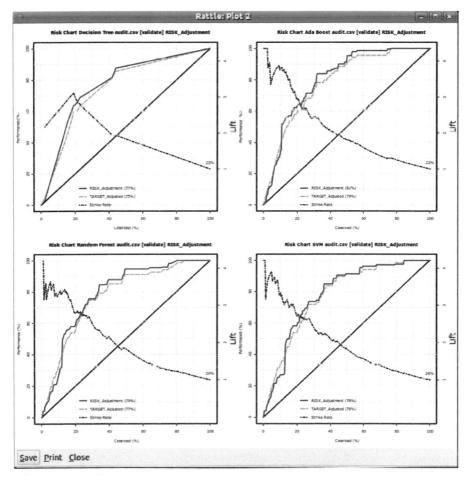

Figure 15.5: Four risk charts displayed to compare performances of multiple model builders on the *audit* dataset.

The plots generated by Rattle include a measure of the area under the curve in the legend of the plot. For Figure 15.4, the area under the target line is 77%. That is, the line splits the plot into two regions, with 77% of the region below the line and 23% of the region above the line.

15.5 ROC Charts

Another set of common performance charts used in data mining are the ROC chart, sensitivity/specificity chart, lift chart, and precision/recall chart. The ROC is probably the most popular, but each can provide some different insights based essentially on the same performance data. The ROC has a form and interpretation similar to the risk chart, though it plots different measures on the axes. **ROCR** is used by Rattle to generate these charts.

We can differentiate each chart by the measures used on the two axes. An ROC chart plots the true positive rate against the false positive rate. The sensitivity/specificity chart plots the true positive rate against the true negative rate. The lift chart plots the relative increase in predictive performance against the rate of positive predictions. The precision/recall chart plots the proportion of true positives out of the positive predictions against the true positive rate.

15.6 Other Charts

A variety of other charts are also supported by Rattle. Some are experimental and implemented directly within Rattle rather than being available in other packages. This includes the Hand chart, which plots a number of measures proposed by David Hand, a senior statistician with a lot of credibility in the data mining community. Cost curves are another measure with quite a long history but have not become particularly popular.

The other performance chart, the Pr v Ob chart, is suitable for evaluating the performance of regression models. Such models predict a continuous outcome, like the dollar amount of a risk, rather than whether or not there is a risk. The Pr v Ob option produces a plot of the actual or observed values on the x-axis with the model-predicted values on the y-axis. A linear model is also fit to the predicted value, based on the actual value, and is displayed in the plot generated by Rattle. A diagonal line (predicted=observed) provides a benchmark as the perfect model (i.e., perfect correlation between the predicted values and the observed values).

15.7 Scoring

The Evaluate tab also provides a Score option. Rather than running a model over some observations and generating a performance measure, the score option allows the predictions to be saved to a file so that we can perform our own analyses of the model using any other tools. Also, it allows us to effectively deploy the model so that we might score new data, save it to a file, and forward the results to the appropriate teams.

Rattle's Score option of the Evaluate tab, when Executed, will apply the selected model(s) to the specified dataset (training/validation/testing/full/CSV file/R dataset), saving the results to a CSV file. Once run, the scores (either the class, probability, or predicted value), together with any Ident variables (or optionally all variables), will be saved to a CSV file for action or further processing with other tools, as desired. A window will pop up to indicate the location and name of the file to which the scores have been saved.

The dataset that is scored must have exactly the same format as the dataset loaded for training the model. Rattle assumes the columns will be in the same order, and we might expect them to have the same names (noting that R is case-sensitive).

Chapter 16

Deployment

Once a model is developed and evaluated, and we have determined it to be suitable, we then need to deploy it. This is an often-overlooked issue in many data mining projects. It also seems to receive little attention when setting up a data mining capability within an organisation. Yet it is an important issue, as we need to ensure we obtain the benefit from the model.

In this chapter, we briefly consider a number of deployment options. We begin with considering a deployment in R. We also consider the conversion of our models into the Predictive Modelling Markup Language (PMML). This allows us to export our model to other systems, which includes systems that can convert the model into C code that can run as a stand-alone module.

16.1 Deploying an R Model

A simple approach to deployment is to use `predict()` to apply the model to a new dataset. We often refer to this as "scoring." Rattle's evaluation tab supports scoring with the **Score** option of the **Evaluate** tab. There are a number of options available. The first is whether to score the training dataset, the validation dataset, the testing dataset, or some dataset loaded from a CSV file (which must contain the exact same variables). Any number of models can be selected, and the results (either as predicted values or probabilities) are written to a CSV file.

Often, we will want to score a new dataset regularly as new observations become available. In this case, we will save the model for later use.

The Rattle concept of a project, as discussed in Section 2.8, is useful in such a circumstance. This will save the current state of Rattle (including the actual data and models built during a session). At a later time, this project can be loaded into a new instance of Rattle (running on the same host or even a different host and operating system). A new dataset can then be scored using the saved model. To do so, we do not need to start up the Rattle GUI but simply access the relevant model (e.g., crs$rf) and apply it to some new data using predict().

As we see in Rattle's Log tab, below the surface, when we save and load projects, we are simply using save() and load(). These create a binary representation of the R objects, saving them to a file and then loading them into R.

A Rattle project can get quite large, particularly with large datasets. Larger files take longer to load, and for deploying a model it is often not necessary to keep the original data. So as we get serious about deployment, we might save just the model we wish to deploy. This is done using save() to save the actual model.

After building a random forest model in Rattle using the *weather* dataset, we might save the model to a file by using save() in the R Console:

```
> myrf <- crs$rf
> save(myrf, file="model01_110501.RData")
```

A warning message may be shown just to suggest that when reloading the binary file into a new instance of R, **rattle** might not have been loaded, and it is perhaps a good idea to do so. This is to ensure the objects that are saved are correctly interpreted by R.

We now want to simulate the application of the model to a new dataset. To do so, we might simply save the current dataset to a CSV file using write.csv():

```
> write.csv(crs$dataset, file="cases_110601.csv")
```

We can then load the model into a different instance of R at a later time using load() and apply (i.e., use predict() on) the model (using a script based on the commands shown in Rattle's Log tab) to a new dataset. In this case, we then also write the results to another CSV file using write.csv():

```
> library(randomForest)
> (load("model01_110501.RData"))

[1] "myrf"

> dataset <- read.csv("cases_110601.csv")
> pr <- predict(myrf, dataset, type="prob")[,2]
> write.csv(cbind(dataset, pr),
            file="scores_110601.csv",
            row.names=FALSE)
> head(cbind(Actual=dataset$TARGET_Adjusted, Predicted=pr))

  Predicted
1     0.688
2     0.712
3     0.916
4     0.164
5     0.052
6     0.016
```

We can see that the random forest model is doing okay on these few observations.

In practise, once model deployment has been approved, the model is deployed into a secure environment. We can then schedule the model to be applied regularly to a new dataset using a script that is very similar to that above. The dataset can be obtained from a data warehouse, for example, and the results populated back into the data warehouse. Other processes might then come into play to make use of the scores, perhaps to identify clients who need to be audited or to communicate to the weather forecaster the predicted likelihood of rain tomorrow.

16.2 Converting to PMML

An alternative approach to direct deployment within R is to export the model in some form so that it might be imported into other software for predictions on new data. Exporting a model to an open standard format facilitates this process. A model represented using this open standard representation can then be exported to a variety of other software or languages.

The Predictive Model Markup Language (Guazzelli et al., 2010, 2009) (PMML) provides such a standard language for representing data mining models. PMML is an XML-based standard that is supported, to some extent, by the major commercial data mining vendors and many open source data mining tools.

The **pmml** package for R is responsible for exporting models from R to PMML. PMML models generated by Rattle using pmml() can be imported into a number of other products. The Export button (whilst displaying a model within the Model tab) will export a model as PMML.

We illustrate here the form that the PMML export takes. First, we again create a dataset object:

```
> library(rattle)
> weatherDS <- new.env()
> evalq({
    data <- weather
    nobs <- nrow(data)
    vars <- -grep('^(Date|Locat|RISK)', names(weather))
    set.seed(42)
    train <- sample(nobs, 0.7*nobs)
    form <- formula(RainTomorrow ~ .)
  }, weatherDS)
```

Next, we build the decision tree model:

```
> library(rpart)
> weatherRPART <- new.env(parent=weatherDS)
> evalq({
    model <- rpart(formula=form, data=data[train, vars])
  }, weatherRPART)
```

Now we can generate the PMML representation using pmml(). We then print some rows from the PMML representation of the model:

```
> library(pmml)
> p <- pmml(weatherRPART$model)
> r <- c(1:4, 7, 12, 35, 36, 69, 71, 137:139)
> cat(paste(strsplit(toString(p), "\n")[[1]][r],
            collapse="\n"))

<PMML version="3.2" xmlns="http://www.dmg.org/PMML-3_2"
    xmlns=...>
 <Header copyright="Copyright (c) 2011 gjw"
     description="RPart Decision Tree Model">
  <Extension name="user" value="gjw" extender="Rattle"/>
  <Application name="Rattle/PMML" version="1.2.27"/>
 <DataDictionary numberOfFields="21">
  <DataField name="MinTemp" optype="continuous" .../>
 </DataDictionary>
 <TreeModel modelName="RPart_Model" ...>
   <Node id="2" score="No" recordCount="204" ...>
     <SimplePredicate field="Pressure3pm"
     operator="greaterOrEqual" value="1011.9"/>
   </Node>
 </TreeModel>
</PMML>
```

16.3 Command Summary

This chapter has referenced the following R packages, commands, functions, and datasets:

load()	command	Load R objects from a file.
pmml()	function	Convert a model to PMML.
pmml	package	Supports conversion of many models.
predict()	function	Score a dataset using a model.
save()	command	Save R objects to a binary file.

Part IV

Appendices

Appendix A

Installing **Rattle**

Rattle relies on an extensive collection of free and open source software. Some preliminary steps need to be followed in installing it. The latest installation instructions are maintained at `http://rattle.togaware.com`. The instructions cover **Rattle** on GNU/Linux, Microsoft Windows and MacOS/X.

Rattle is distributed as a freely available open source R package available from CRAN, the Comprehensive R Archive Network (`http://cran.r-project.org/`). The source code for **Rattle** is available from Google Code (`http://code.google.com/p/rattle/`). A discussion mailing list is available from Google Groups (`http://groups.google.com/group/rattle-users`).

If you are setting up a new data mining platform, the recommended approach is to build it on top of the Ubuntu operating system (`http://ubuntu.com`). This delivers a free and open source environment for data mining.

If you already have R installed and have installed the appropriate GTK libraries for your operating system, then installing **Rattle** is as simple as:

```
> install.packages("rattle")
```

Once installed, the function `rattleInfo()` provides version information for rattle and dependencies and will also check for available updates and generate the command that can be cut-and-pasted to update the appropriate packages:

```
> rattleInfo()

Rattle: version 2.6.7 cran 2.6.6
R: version 2.13.0 (2011-04-13) (Revision 55427)

Sysname: Linux
Release: 2.6.35-28-generic
Version: #49-Ubuntu SMP Tue Mar 1 14:39:03 UTC 2011
Nodename: nyx
Machine: x86_64
Login: unknown
User: gjw

Installed Dependencies
RGtk2: version 2.20.12
pmml: version 1.2.27
bitops: version 1.0-4.1
colorspace: version 1.1-0
ada: version 2.0-2
amap: version 0.8-5
arules: version 1.0-6
biclust: version 1.0.1
cairoDevice: version 2.15
cba: version 0.2-6
descr: version 0.3.4
doBy: version 4.3.1 upgrade available 4.4.0
e1071: version 1.5-26
ellipse: version 0.3-5
fEcofin: version 290.76
fBasics: version 2110.79
foreign: version 0.8-44
fpc: version 2.0-3
gdata: version 2.8.2
gtools: version 2.6.2
gplots: version 2.8.0
```

```
gWidgetsRGtk2: version 0.0-74
Hmisc: version 3.8-3
kernlab: version 0.9-12
latticist: version 0.9-43
Matrix: version 0.999375-50
mice: version 2.8
network: version 1.6
nnet: version 7.3-1
odfWeave: version 0.7.17
party: version 0.9-99992
playwith: version 0.9-53
psych: version 1.0-97 upgrade available 1.0-98
randomForest: version 4.6-2
RBGL: version 1.28.0
RColorBrewer: version 1.0-2
reshape: version 0.8.4
rggobi: version 2.1.17
RGtk2Extras: version 0.5.0
ROCR: version 1.0-4
RODBC: version 1.3-2
rpart: version 3.1-50
rpart.plot: version 1.2-2
RSvgDevice: version 0.6.4.1
survival: version 2.36-9
timeDate: version 2130.93
XML: version 3.4-0

Upgrade the packages with either:

  > install.packages(c("doBy", "psych"))

  > install.packages(rattleInfo())
```

Appendix B

Sample Datasets

The following sections introduce the datasets that we use throughout the book to demonstrate data mining. R provides quite a collection of datasets. Each of the datasets we introduce here is available through R packages and may also be available from other sources.

In addition to introducing the datasets themselves, we also illustrate how they were derived. This entails presenting many new concepts from the R language. We purposefully do so within a real context of manipulating data to generate data suitable for data mining. A detailed understanding of many of these R programming constructs is not a necessity for understanding the material in this book. You are encouraged, though, to work through these programming examples at some time, as a data miner will need sophisticated tools for manipulating data.

Using the datasets supplied, as we do to demonstrate Rattle, oversimplifies the situation. This data will generally already be in a form suitable for mining, and in reality this is not what we find in a data mining project. In practise the data we are confronted with will come from a multitude of sources and will be in need of a variety of processing steps to clean it up and merge it into a single dataset. We have a lot of work in front of us in transforming a multitude of data into a form suitable for data mining. As we have reinforced throughout this is a major task.

The sizes of the datasets that we use throughout this book (and provided by **rattle**) are summarised below using dim(). The *weather* dataset, for example, has 366 observations and 24 variables:

```
> library(rattle)
> dim(weather)

[1] 366   24

> dim(weatherAUS)

[1] 39996    24

> dim(audit)

[1] 2000    13
```

B.1   Weather

We have seen the *weather* dataset in Chapter 2, where we identified it as being obtained[1] from the Australian Bureau of Meteorology's Web site. This is quite a small dataset, allowing the concepts we cover to be presented concisely. On the surface, it is not too complex, and most of us can relate to the issue of whether it might rain tomorrow! Keep in mind that real-world datasets tend not to be so small, and large datasets present many of the challenges we face in data mining.

The data used here (i.e., the actual *weather* dataset from **rattle**) comes from a weather monitoring station located in Canberra, Australia. The Bureau makes available 13 months of daily weather observations from many locations across Australia. The data is available as CSV (comma-separated value) files. The full list is available from http:// www.bom.gov.au/climate/dwo/. Similar data is available from other government authorities around the world, as well as from many personal weather stations, which are now readily available.

B.1.1   Obtaining Data

The weather data for a specific month can be downloaded within R by using `read.csv()`. The process of doing so illustrates our interaction with R.

---

[1]Permission to use the dataset for this book and associated package (**rattle**) was obtained 17 December 2008 from mailto:webclim@bom.gov.au, and the Australian Bureau of Meteorology is acknowledged as the source of the data.

First, we need to determine the date for which we want to download data. For this, `Sys.Date()`, which returns the current date, will be useful:

```
> Sys.Date()
[1] "2011-06-13"
```

The current date can be passed to `format()`:

```
> today <- format(Sys.Date(), format="%Y%m")
```

This will generate a string consisting of the year and month, as specified by the `format=` argument to the function. The result of `format()` is stored into memory with a reference name of `today`. This is achieved using the assignment operator, `<-`, which itself is a function in R. No output is produced from the command above—the value is saved into the specified memory. We can inspect its contents simply by typing its name:

```
> today
[1] "201106"
```

Now we build up the actual Web address (the URL from which we can download the data) using `paste()`. This function takes a collection of strings and pastes them together into one string. Here we override the default behaviour of `paste()` using the `sep=` (i.e., separator) argument so that there will be no spaces between the strings that we are pasting together. We will save the result of the pasting into a memory reference named `bom`:

```
> bom <- paste("http://www.bom.gov.au/climate/dwo/", today,
               "/text/IDCJDW2801.", today, ".csv", sep="")
```

The string referenced by the name `bom` is then the URL[2] to extract the current month's data for the Canberra weather station (identified as IDCJDW2801):

```
> bom
[1] "http:[...]/dwo/201106/text/IDCJDW2801.201106.csv"
```

---

[2]Note that the URL here is correct at the time of publication but could change over time.

Note the use of the string "[...]" in the output—this is used so that the result is no wider than the printed page. We will use this notation often to indicate where data has been removed for typesetting purposes.

The most recent observations from the Bureau can now be read. Note that, for the benefit of stability, the actual dataset used below is from a specific date, June 2009, and the details of other data obtained at different times will differ. The first few lines of the downloaded file contain information about the location, and so we skip those lines by using the skip= argument of read.csv(). The check.names= argument is set to FALSE (the default is TRUE) so that the column names remain exactly as recorded in the file:

```
> dsw <- read.csv(bom, skip=6, check.names=FALSE)
```

By default, R will convert them into names that it can more easily handle, for example, replacing spaces with a period. We will fix the names ourselves shortly.

The dataset is not too large, as shown by dim() (consisting of up to one month of data), and we can use names() to list the names of the variables included:

```
> dim(dsw)

[1] 28 22

> head(names(dsw))

[1] ""
[2] "Date"
[3] "Minimum temperature (\xb0C)"
[4] "Maximum temperature (\xb0C)"
[5] "Rainfall (mm)"
[6] "Evaporation (mm)"
```

Note that, if you run this code yourself, the dimensions will most likely be different. The data you download today will be different from the data downloaded when this book was processed by R. In fact, the number of rows should be about the same as the day number of the current month.

B.1.2   Data Preprocessing

We do not want all of the variables. In particular, we will ignore the first column and the time of the maximum wind gust (variable number 10). The first command below will remove these two columns from the dataset. We then simplify the names of the variables to make it easier to refer to them. This can be done as follows using names():

```
> ndsw <- dsw[-c(1, 10)]
> names(ndsw) <- c("Date", "MinTemp", "MaxTemp",
    "Rainfall", "Evaporation", "Sunshine",
    "WindGustDir", "WindGustSpeed", "Temp9am",
    "Humidity9am", "Cloud9am", "WindDir9am",
    "WindSpeed9am", "Pressure9am", "Temp3pm",
    "Humidity3pm", "Cloud3pm", "WindDir3pm",
    "WindSpeed3pm", "Pressure3pm")
```

We can now check that the new dataset has the right dimensions and variable names:

```
> dim(ndsw)

[1] 28 20

> names(ndsw)

 [1] "Date"          "MinTemp"         "MaxTemp"
 [4] "Rainfall"      "Evaporation"     "Sunshine"
 [7] "WindGustDir"   "WindGustSpeed"   "Temp9am"
[10] "Humidity9am"   "Cloud9am"        "WindDir9am"
[13] "WindSpeed9am"  "Pressure9am"     "Temp3pm"
[16] "Humidity3pm"   "Cloud3pm"        "WindDir3pm"
[19] "WindSpeed3pm"  "Pressure3pm"
```

B.1.3   Data Cleaning

We must also clean up some of the variables. We start with the wind speed. To view the first few observations, we can use head(). We further limit our review to just three variables, which we explicitly list as the column index:

```
> vars <- c("WindGustSpeed","WindSpeed9am","WindSpeed3pm")
> head(ndsw[vars])

  WindGustSpeed WindSpeed9am WindSpeed3pm
1            24         Calm           15
2            31            6           13
3            22            9           17
4            11            6         Calm
5            20            6            7
6            39            2           28
```

Immediately, we notice that not all the wind speeds are numeric. The variable WindSpeed9am has a value of *Calm* for the first observation, and so R is representing this data as a categoric, and not as a numeric as we might be expecting. We can confirm this using class() to tell us what class of data type the variable is.

First, we confirm that **ndsw** is a data frame (which is R's representation of a dataset):

```
> class(ndsw)

[1] "data.frame"
```

With the next example, we introduce apply() to apply class() to each of the variables of interest. We confirm that the variables are character strings:

```
> apply(ndsw[vars], MARGIN=2, FUN=class)

WindGustSpeed  WindSpeed9am  WindSpeed3pm
  "character"   "character"   "character"
```

The MARGIN= argument chooses between applying the supplied function to the rows of the dataset or to the columns (i.e., variables) of the dataset. The 2 selects columns, whilst 1 selects rows. The function that is applied is supplied with the FUN= argument.

To transform these variables, we introduce a number of common R constructs. We first ensure that we are treating the variable as a character string by converting it (although somewhat redundantly in this case) with as.character():

```
> ndsw$WindSpeed9am   <- as.character(ndsw$WindSpeed9am)
> ndsw$WindSpeed3pm   <- as.character(ndsw$WindSpeed3pm)
> ndsw$WindGustSpeed  <- as.character(ndsw$WindGustSpeed)
> head(ndsw[vars])

  WindGustSpeed WindSpeed9am WindSpeed3pm
1            24         Calm           15
2            31            6           13
3            22            9           17
4            11            6         Calm
5            20            6            7
6            39            2           28
```

B.1.4 Missing Values

We next identify that empty values (i.e., an empty string) represent missing data, and so we replace them with R's notion of missing values (NA). The within() function can be used to allow us to directly reference variables within the dataset without having to prefix them with the name of the dataset (i.e., avoiding having to use ndsw$WindSpeed9am):

```
> ndsw <- within(ndsw,
         {
             WindSpeed9am[WindSpeed9am   == ""] <- NA
             WindSpeed3pm[WindSpeed3pm   == ""] <- NA
             WindGustSpeed[WindGustSpeed == ""] <- NA
         })
```

Then, *Calm*, meaning no wind, is replaced with 0, which suits our numeric data type better:

```
> ndsw <- within(ndsw,
         {
             WindSpeed9am[WindSpeed9am   == "Calm"] <- "0"
             WindSpeed3pm[WindSpeed3pm   == "Calm"] <- "0"
             WindGustSpeed[WindGustSpeed == "Calm"] <- "0"
         }) .
```

Finally, we convert the character strings to the numbers they actually represent using as.numeric(), and check the data type class to confirm they are now numeric:

```
> ndsw <- within(ndsw,
        {
            WindSpeed9am  <- as.numeric(WindSpeed9am)
            WindSpeed3pm  <- as.numeric(WindSpeed3pm)
            WindGustSpeed <- as.numeric(WindGustSpeed)
        })
> apply(ndsw[vars], 2, class)

WindGustSpeed  WindSpeed9am  WindSpeed3pm
    "numeric"     "numeric"     "numeric"
```

The wind direction variables also need some transformation. We see below that the wind direction variables are categoric variables (they are technically *factors* in R's nomenclature). Also note that one of the possible values is the string consisting of just a space, and that the levels are ordered alphabetically:

```
> vars <- c("WindSpeed9am","WindSpeed3pm","WindGustSpeed")
> head(ndsw[vars])

  WindSpeed9am WindSpeed3pm WindGustSpeed
1            0           15            24
2            6           13            31
3            9           17            22
4            6            0            11
5            6            7            20
6            2           28            39

> apply(ndsw[vars], 2, class)

 WindSpeed9am  WindSpeed3pm WindGustSpeed
    "numeric"     "numeric"     "numeric"

> levels(ndsw$WindDir9am)

 [1] " "    "E"    "ENE" "ESE" "N"    "NNW" "NW"  "S"    "SE"
[10] "SSE" "SW"   "WNW"
```

To deal with missing values, which are represented in the data as an empty string (corresponding to a wind speed of zero), we map such data to NA:

```
> ndsw <- within(ndsw,
  {
    WindDir9am[WindDir9am == " "] <- NA
    WindDir9am[is.na(WindSpeed9am) |
               (WindSpeed9am == 0)] <- NA

    WindDir3pm[WindDir3pm == " "] <- NA
    WindDir3pm[is.na(WindSpeed3pm) |
               (WindSpeed3pm == 0)] <- NA

    WindGustDir[WindGustDir == " "] <- NA
    WindGustDir[is.na(WindGustSpeed) |
                (WindGustSpeed == 0)] <- NA
  })
```

B.1.5   Data Transforms

Another common operation on a dataset is to create a new variable from other variables. An example is to capture whether it rained today. This can be simply determined, by definition, through checking whether there was more than 1 mm of rain today. We use ifelse() to do this in one step:

```
> ndsw$RainToday <- ifelse(ndsw$Rainfall > 1, "Yes", "No")
> vars <- c("Rainfall", "RainToday")
> head(ndsw[vars])
  Rainfall RainToday
1     0.6        No
2     0.0        No
3     1.6       Yes
4     8.6       Yes
5     2.2       Yes
6     1.4       Yes
```

We want to also capture and associate with today's observation whether it rains tomorrow. This is to become our target variable. Once again, if it rains less than 1 mm tomorrow, then we report that as no rain. To capture this variable, we need to consider the observation of rainfall recorded on the following day. Thus, when we are considering today's observation (e.g., observation number 1), we want to consider tomorrow's observation (observation 2) of `Rainfall`. That is, there is a lag of one day in determining today's value of the variable `RainTomorrow`.

A simple approach can be used to calculate `RainTomorrow`. We simply note the value of `RainToday` for the next day's observation. Thus, we build up the list of observations for `RainToday` starting with the second observation (ignoring the first). An additional observation then needs to be added for the final day (for which we have no observation for the following day):

```
> ndsw$RainTomorrow <- c(ndsw$RainToday[2:nrow(ndsw)], NA)
> vars <- c("Rainfall", "RainToday", "RainTomorrow")
> head(ndsw[vars])

  Rainfall RainToday RainTomorrow
1      0.6        No           No
2      0.0        No          Yes
3      1.6       Yes          Yes
4      8.6       Yes          Yes
5      2.2       Yes          Yes
6      1.4       Yes           No
```

Finally, we would also like to record the amount of rain observed "tomorrow." This is achieved as follows using the same lag approach:

```
> ndsw$RISK_MM <- c(ndsw$Rainfall[2:nrow(ndsw)], NA)
> vars <- c("Rainfall", "RainToday",
            "RainTomorrow", "RISK_MM")
> head(ndsw[vars])

  Rainfall RainToday RainTomorrow RISK_MM
1      0.6        No           No     0.0
2      0.0        No          Yes     1.6
3      1.6       Yes          Yes     8.6
4      8.6       Yes          Yes     2.2
5      2.2       Yes          Yes     1.4
6      1.4       Yes           No     0.8
```

The source dataset has now been processed to include a variable that we might like to treat as a target variable—to indicate whether it rained the following day.

B.1.6 Using the Data

Using this historic data, we can now build a model (as we did in Chapter 2) that might help us to decide whether we need to take an umbrella with us tomorrow if we live in Canberra. (You may like to try this on local data for your own region.)

Above, we retrieved up to one month of observations. We can repeat the process, using the same code, to obtain 12 months of observations. This has been done to generate the *weather* dataset provided by Rattle. The *weather* dataset covers only Canberra for the 12 month period from 1 November 2007 to 31 October 2008 inclusive.

Rattle also provides the *weatherAUS* dataset, which captures the weather observations for more than a year from over 45 weather observation stations throughout Australia. The format of the *weatherAUS* dataset is exactly the same as for the *weather* dataset. In fact, the *weather* dataset is a subset of the *weatherAUS* dataset, and we could reconstruct it with the following R code using subset():

```
> cbr <- subset(weatherAUS,
                Location == "Canberra" &
                Date >= "2007-11-01" &
                Date <= "2008-10-31")
```

The subset() function takes as its first argument a dataset, and as its second argument a logical expression that specifies the rows of the data that we wish to retain in the result.

We can check that this results in the same dataset as the *weather* dataset by simply testing if they are equal using ==:

```
> cbr == weather
```

This will print a lot of TRUEs to the screen, as it compares each value from the *cbr* dataset with the corresponding value from the *weather* dataset. We could have a look through what is printed to make sure they are all TRUE, but that's not very efficient.

Instead, we can find the number of actual values that are compared. First, we get the two dimensions, using dim(), and indeed the two datasets have the same dimensions:

```
> dim(cbr)
[1] 366   24
> dim(weather)
[1] 366   24
```

Then we calculate the number of data items within the dataset. To do this, we could multiply the first and second values returned by `dim()`. Instead, we will introduce two new handy functions to return the number of rows and the number of columns in the dataset:

```
> dim(cbr)[1] * dim(cbr)[2]
[1] 8784
> nrow(cbr) * ncol(cbr)
[1] 8784
```

Now we count the number of TRUEs when comparing the two datasets value by value. Noting that TRUE corresponds to the numeric value 1 in R and FALSE corresponds to 0, we can simply sum the data. We need to remove NAs to get a sum, otherwise `sum()` will return NA. We also need to count the number of NAs removed, which we do. Note that the totals all add up to 8784! The two datasets are the same.

```
> sum(cbr == weather, na.rm=TRUE)
[1] 8737
> sum(is.na(cbr))
[1] 47
> sum(is.na(weather))
[1] 47
> sum(cbr == weather, na.rm=TRUE) + sum(is.na(cbr))
[1] 8784
```

The sample *weather* dataset can also be downloaded directly from the **Rattle** Web site:

```
> twweather <- "http://rattle.togaware.com/weather.csv"
> myweather <- read.csv(twweather)
```

B.2 Audit

Another dataset we will use for illustrating data mining is the *audit* dataset, which is also provided by **rattle**. The data is artificial but reflects a real-world dataset used for reviewing the outcomes of historical financial audits. Picture, for example, your country's revenue authority (e.g., the Internal Revenue Service in the USA, Inland Revenue in the UK or the Australian Taxation Office). Revenue authorities collect taxes and reconcile the taxes we pay each year against the information we supply to them.

Many thousands of audits of taxpayers' tax returns might be performed each year. The outcome of an audit may be productive, in which case an adjustment to the information supplied was required, usually resulting in a change to the amount of tax that the taxpayer is liable to pay (an increase or a decrease). An unproductive audit is one for which no adjustment was required after reviewing the taxpayer's affairs.

The *audit* dataset attempts to simulate this scenario. It is supplied as both an R data file and a CSV file. The dataset consists of 2000 fictional taxpayers who have been audited for tax compliance. For each case, an outcome of the audit is recorded (i.e., whether the financial claims had to be adjusted or not). The actual dollar amount of any adjustment that resulted is also recorded (noting that adjustments can go in either direction). The *audit* dataset contains 13 variables, with the first variable being a unique client identifier. It is, in fact, derived from the so-called *adult* dataset.

B.2.1 The Adult Survey Dataset

Like the *weather* dataset, the *audit* dataset is actually derived from another freely available dataset. Unlike the *weather* dataset, the *audit* dataset is purely fictional. We will discuss here how the data was derived from the so-called adult *survey* dataset available from the University of California at Irvine's machine learning[3] repository. We use this dataset as the starting point and will perform various transformations of the data with the aim of building a dataset that looks more like an audit dataset. With the purpose of further illustrating the data manipulation capabilities of R, we review the derivation process.

---

[3]http://archive.ics.uci.edu/ml/.

First, we `paste()` together the constituent parts of the URL from which the dataset is obtained. Note the use of the `sep=` (separator) argument to include a "/" between the constituent parts. We then use `download.file()` to retrieve the actual file from the Internet, and to save it under the name `survey.csv`:

```
> uci <- paste("ftp://ftp.ics.uci.edu/pub",
              "machine-learning-databases",
              "adult/adult.data", sep="/")
> download.file(uci, "survey.csv")
```

The file is now stored locally and can be loaded into R. Because the file is a CSV file, we use `read.csv()` to load it into R:

```
> survey <- read.csv("survey.csv", header=FALSE,
                    strip.white=TRUE, na.strings="?",
              col.names=c("Age", "Workclass", "fnlwgt",
                  "Education", "Education.Num",
                  "Marital.Status", "Occupation",
                  "Relationship", "Race", "Gender",
                  "Capital.Gain", "Capital.Loss",
                  "Hours.Per.Week", "Native.Country",
                  "Salary.Group"))
```

The additional arguments of `read.csv()` are used to fine-tune how the data is read into R. The `header=` argument needs to be set to `FALSE` since the data file has no header row (a header row is the first row and lists the variable or column names—here, though, it is data, not a header). We set `strip.white=` to `TRUE` to strip spaces from the data to ensure we do not get extra white space in any columns. Missing values are notated with a question mark, so we tell the function this with the `na.strings=` argument. Finally, we supply a list of variable names using the `col.names=` (column names) argument.

B.2.2   From Survey to Audit

We begin to turn the *survey* data into the *audit* dataset, first by selecting a subset of the columns and then renaming some of the columns (reinforcing again that this is a fictitious dataset):

```
> audit <- survey[,c(1:2,4,6:8,10,12:14,11,15)]
> names(audit)[c(seq(2, 8, 2), 9:12)] <-
    c("Employment", "Marital", "Income", "Deductions",
      "Hours", "Accounts", "Adjustment", "Adjusted")
```

Here we see a couple of interesting language features of R. We have previously seen the use of names() to retrieve the variable names from the dataset. This function returns a list of names, and we can index the data items within the list as usual. The interesting feature here is that we are assigning into that resulting list another list. The end result is that the variable names are thereby actually changed within the dataset:

```
> names(audit)

 [1] "Age"        "Employment" "Education"  "Marital"
 [5] "Occupation" "Income"     "Gender"     "Deductions"
 [9] "Hours"      "Accounts"   "Adjustment" "Adjusted"
```

B.2.3 Generating Targets

We now look at what will become the output variables, Adjustment and Adjusted. These will be interpreted as the dollar amount of any adjustment made to the tax return and whether or not there was an adjustment made, respectively. Of course, they need to be synchronised.

The variable Adjusted is going to be a binary integer variable that takes on the value 0 when an audit was not productive and the value 1 when an audit was productive. Initially the variable is a categoric variable (i.e., of class *factor* in R's nomenclature) with two distinct values (i.e., two distinct *levels* in R's nomenclature). We use R's table() to report on the number of observations having each of the two distinct output values:

```
> class(audit$Adjusted)

[1] "factor"

> levels(audit$Adjusted)

[1] "<=50K" ">50K"

> table(audit$Adjusted)

<=50K  >50K
24720  7841
```

We now convert this into a binary integer variable for convenience. This is not strictly necessary, but often our mathematics in describing algorithms works nicely when we think of the target as being 0 or 1. The `as.integer()` function will transform a categoric variable into a numeric variable. R uses the integer 1 to represent the first categoric value and 2 to represent the second categoric value. So to turn this into our desired 0 and 1 values we simply subtract 1 from each integer:

```
> audit$Adjusted <- as.integer(audit$Adjusted)-1
> class(audit$Adjusted)

[1] "numeric"

> table(audit$Adjusted)

    0     1
24720  7841
```

It is instructive to understand the subtraction that is performed here. In particular, the 1 is subtracted from each data item. In R, we can subtract one list of numbers from another, but they generally need to be the same length. The subtraction occurs pairwise. If one list is shorter than the other, then it is recycled as many times as required to perform the operation. Thus, 1 is recycled as many times as the number of observations of `Adjusted`, with the end result we noted. The concept can be illustrated simply:

```
> 11:20 - 1:10

 [1] 10 10 10 10 10 10 10 10 10 10

> 11:20 - 1:5

 [1] 10 10 10 10 10 15 15 15 15 15

> 11:20 - 1

 [1] 10 11 12 13 14 15 16 17 18 19
```

Some mathematics is now required to ensure that most productive cases (those observations for which `Adjusted` is 1) have an adjustment (i.e., the variable `Adjustment` is nonzero) and nonproductive cases necessarily have an adjustment of 0.

We first calculate the number of productive cases that have a zero adjustment (saving the result into the reference `prod`) and the number of nonproductive cases that have a nonzero adjustment (saving the result into the reference nonp):

```
> prod <- sum(audit$Adjusted == 1 & audit$Adjustment == 0)
> prod

[1] 6164

> nonp <- sum(audit$Adjusted == 0 & audit$Adjustment != 0)
> nonp

[1] 1035
```

This example again introduces a number of new concepts from the R language. We will break them down one at a time and then come back to the main story.

Recall that the notation audit$Adjusted refers to the observations of the variable Adjusted of the *audit* dataset. As with the subtraction of a single value, 1, from such a list of observations, as we saw above, the comparison operator == (as well as != to test not equal) operates over such data. It tests each observation to see if it is equal to, for example, 1.

The following example illustrates this. Consider just the first few observations of the variables Adjusted and Adjustment. R notates logical variables with the observations TRUE or FALSE. The "&" operator is used for comparing lists of logical values pairwise:

```
> obs <- 1:9
> audit$Adjusted[obs]

[1] 0 0 0 0 0 0 0 1 1

> audit$Adjusted[obs]==1

[1] FALSE FALSE FALSE FALSE FALSE FALSE FALSE  TRUE  TRUE

> audit$Adjustment[obs]

[1]  2174    0    0    0    0    0    0    0 14084

> audit$Adjustment[obs] == 0

[1] FALSE  TRUE  TRUE  TRUE  TRUE  TRUE  TRUE  TRUE FALSE

> audit$Adjusted[obs] == 1 & audit$Adjustment[obs] == 0

[1] FALSE FALSE FALSE FALSE FALSE FALSE FALSE  TRUE FALSE
```

Notice that the final example here is used as the argument to sum() in our code above. This summation relies on the fact that TRUE is treated as the integer 1 and FALSE as the integer 0 when needed. Thus

```
> sum(audit$Adjusted[obs]==1 & audit$Adjustment[obs]==0)
[1] 1
```

so that there is only one observation where both of the conditions are TRUE. Over the whole dataset:

```
> sum(audit$Adjusted == 1 & audit$Adjustment == 0)
[1] 6164
```

For these 6164 observations, we note that they are regarded as being adjusted yet there is no adjustment amount recorded for them. So, for some majority of these observations, we want to ensure that they do have an adjustment recorded for them, as we would expect from real data.

The following formulation uses the integer division function %/% to divide prod by nonp and then multiply the result by nonp. This will usually result in a number that is a little less than the original prod and will be the number of observations that we will adjust to have a nonzero Adjustment:

```
> adj <- (prod %/% nonp) * nonp
```

The resulting value, saved as adj (5175), is thus an integer multiple (5) of the value of nonp (1035). The significance of this will be apparent shortly.

Now we make the actual change from 0 to a random integer. To do so, we take the values that are present in the data for adjustments where Adjusted is actually 0 (the nonp observations) and multiply them by a random number. The result is assigned to an adjusted observation that currently has a 0 Adjustment. The point around the integer multiple of nonp, noted above, is that the following will effectively replicate the current nonp observations an integer number of times to assign them to the subset of the prod observations being modified:

```
> set.seed(12345)
> audit[audit$Adjusted == 1 & audit$Adjustment == 0,
        'Adjustment'][sample(prod, adj)] <-
    as.integer(audit[audit$Adjusted == 0 &
                      audit$Adjustment != 0, 'Adjustment'] *
              rnorm(adj, 2))
```

There is quite a lot happening in these few lines of code. First off, because we are performing random sampling (using sample() and rnorm()), we use set.seed() to start the random number generation from a known point, and thus the process is repeatable (we will get the same random numbers each time). We could have used any number as the argument to set.seed() as long as we use the same number each time.

Next, we see quite a complex assignment. First, we index *audit* to include just those observations marked as adjusted yet having no adjustment. For these observations, we extract just the Adjustment variable (noting that all resulting observations will be 0). The point of the expression, though, is to identify the locations in memory where this data is stored.

The variable is further indexed by a random sample (using sample()) of adj (5175) numbers between 1 and prod (6164). These are the observations for which we will be changing the Adjustment from 0 to something else.

The remainder of the assignment command calculates the replacement numbers. This time *audit* is indexed to obtain those nonproductive observations with a nonzero adjustment. These 5175 values are multiplied by a sequence of adj (5175) random numbers. The random numbers are normally distributed with a mean of 2 and are generated using rnorm().

That is quite a complex operation on the data. With a little bit of familiarity, and breaking down the operation into its constituent parts, we can understand what it does.

We need to tidy up one last operation involving the Adjustment variable. Observations marked as having a nonproductive outcome should have a value of 0 for Adjustment. The following will achieve this:

```
> audit[audit$Adjusted == 0 & audit$Adjustment != 0,
        'Adjustment'] <- 0
```

B.2.4   Finalising the Data

The remainder of the operations we perform on the *audit* dataset are similar in nature, and we now finalise the data. The observations of Deductions, for nonadjusted cases, are reduced to be closer to 0, reflecting a likely scenario:

```
> audit[audit$Adjusted==0, 'Deductions'] <-
    audit[audit$Adjusted==0, 'Deductions']/1.5
```

To keep to a smaller dataset for illustrative purposes, we sample 2000 rows:

```
> set.seed(12345)
> cases <- sample(nrow(audit), 2000)
```

Finally, we add to the beginning of the variables contained in the dataset a new variable that serves the role of a unique identifier. The identifiers are randomly generated using runif(). This generates random numbers from a uniform distribution. We use it to generate 2000 random numbers of seven digits:

```
> set.seed(12345)
> idents <- as.integer(sort(runif(2000, 1000000, 9999999)))
> audit <- cbind(ID=idents, audit[cases,])
```

B.2.5   Using the Data

The final version of the *audit* dataset, as well as being available from **rattle**, can also be downloaded directly from the Rattle Web site:

```
> twaudit <- "http://rattle.togaware.com/audit.csv"
> myaudit <- read.csv(twaudit)
```

B.3   Command Summary

This appendix has referenced the following R packages, commands, functions, and datasets:

apply()	function	Apply a function over a list.
as.character()	function	Convert to character string.
as.integer()	function	Convert to integer.
as.numeric()	function	Convert to numeric.
audit	dataset	Sample dataset from **rattle**.
class()	function	Identify type of object.
dim()	function	Report the rows/columns of a dataset.
download.file()	function	Download file from URL.
format()	function	Format an object.
head()	function	Show top observations of a dataset.
names()	function	Show variables contained in a dataset.
paste()	function	Combine strings into one string.
read.csv()	function	Read a comma-separated data file.
rnorm()	function	Generate random numbers.
sample()	function	Generate a random sample of numbers.
subset()	function	Create a subset of a dataset.
sum()	function	Add the supplied numbers.
survey	dataset	A sample dataset from UCI repository.
Sys.Date()	function	Return the current date and time.
table()	function	Summarise distribution of a variable.
weather	dataset	Sample dataset from **rattle**.
weatherAUS	dataset	A larger dataset from **rattle**.
within()	function	Perform actions within a dataset.

References

Numbers at the end of each reference indicate the page(s) on which the work is discussed in the text.

Adler, J. (2010), *R in a Nutshell: A Desktop Reference Guide*, O'Reilly, Sebastopol. ix

Amaratunga, D., Cabrera, J., and Lee, Y.-S. (2008), Enriched random forests, *Bioinformatics* **24**(18), 2010–2014. 264

Andrews, F. (2010), *latticist: A Graphical User Interface for Exploratory Visualisation*. R package version 0.9-43. http://latticist.googlecode.com/. 31, 137, 138

Bauer, E. and Kohavi, R. (1999), An empirical comparison of voting classification algorithms: Bagging, boosting, and variants, *Machine Learning* **36**(1–2), 105–139. http://citeseer.ist.psu.edu/bauer99empirical.html. 285

Bivand, R. S., Pebesma, E. J., and Gómez-Rubio, V. (2008), *Applied Spatial Data Analysis with R*, Use R!, Springer, New York. 18

Breiman, L. (1996), Bagging predictors, *Machine Learning* **24**(2), 123–140. http://citeseer.ist.psu.edu/breiman96bagging.html. 264

Breiman, L. (2001), Random forests, *Machine Learning* **45**(1), 5–32. 264

Breiman, L., Friedman, J., Olshen, R., and Stone, C. (1984), *Classification and Regression Trees*, Wadsworth and Brooks, Monterey, CA. 215

Chambers, J. M. (2008), *Software for Data Analysis: Programming with R*, Springer, New York. http://stat.stanford.edu/~jmc4/Rbook/. 68

Cleveland, W. S. (1993), *Visualizing Data*, Hobart Press, Summit, NJ. 108

Cook, D. and Swayne, D. F. (2007), *Interactive and Dynamic Graphics for Data Analysis*, Springer, New York. 138, 141, 148

Cowpertwait, P. S. P. and Metcalfe, A. V. (2009), *Introductory Time Series with R*, Springer, New York. 18

Crano, W. D. and Brewer, M. B. (2002), *Principles and Methods of Social Research*, Lawrence Erlbaum Associates, Mahwah, NJ. 176

CRISP-DM (1996), *Cross Industry Process—Data Mining.* http://www.crisp-dm.org/. 7

Culp, M., Johnson, K., and Michailidis, G. (2010), *ada: An R Package for Stochastic Boosting.* R package version 2.0-2. http://CRAN.R-project.org/package=ada. 275

Cypher, A., ed. (1993), *Watch What I Do: Programming by Demonstration*, The MIT Press, Cambridge, MA. http://www.acypher.com/wwid/WWIDToC.html. 11

Dalgaard, P. (2008), *Introductory Statistics with R*, 2nd ed., Statistics and Computing, Springer, New York. ix, 18

DebRoy, S. and Bivand, R. (2011), *foreign: Read Data Stored by Minitab, S, SAS, SPSS, Stata, Systat, and dBase.* R package version 0.8-44. http://CRAN.R-project.org/package=foreign. 87

Durtschi, C., Hillison, W., and Pacini, C. (2004), The effective use of Benford's law to assist in detecting fraud in accounting data, *Journal of Forensic Accounting* **5**, 17–34. 119

Fraley, C. and Raftery, A. E. (2006), *MCLUST Version 3 for R: Normal Mixture Modeling and Model-Based Clustering*, Technical Report 504, Department of Statistics, University of Washington. 190

Freund, Y. and Mason, L. (1997), The alternating decision tree learning algorithm, in *Proceedings of the 16th International Conference on Machine Learning (ICML99)*, Bled, Slovenia, Morgan Kaufmann, San Fransisco, CA, pp. 124–133. 286

Freund, Y. and Schapire, R. E. (1995), A decision-theoretic generalization of on-line learning and an application to boosting, in P. M. B. Vitányi, ed., *Proceedings of the 2nd European Conference on Computational Learning Theory (EuroCOLT95)*, Barcelona, Spain, Vol. 904 of *Lecture Notes in Computer Science*, Springer, pp. 23–37. http://citeseer.ist.psu.edu/freund95decisiontheoretic.html. 285

Galton, F. (1885), Regression towards mediocrity in hereditary stature, *Journal of the Anthropological Institute* **15**, 246–263. 176

Glade (1998), *A User Interface Designer*, open source. http://glade.gnome.org. xii

Gnome (1997), *The Free Software Desktop Project*, open source. http://glade.gnome.org/. xii

Guazzelli, A., Lin, W.-C., and Jena, T. (2010), *PMML in Action*, CreateSpace, Charleston, SC. 325

Guazzelli, A., Zeller, M., Lin, W.-C., and Williams, G. (2009), PMML: An open standard for sharing models, *The R Journal* **1**(1), 60–65. http://journal.r-project.org/2009-1/RJournal_2009-1_Guazzelli+et+al.pdf. 325

Hahsler, M., Buchta, C., Gruen, B., and Hornik, K. (2011), *arules: Mining Association Rules and Frequent Itemsets*. R package version 1.0-6. http://CRAN.R-project.org/. 201

Han, J. and Kamber, M. (2006), *Data Mining: Concepts and Techniques*, 2nd ed., Morgan Kaufmann, San Francisco, CA. 19

Harrell, Jr., F. E. (2010), *Hmisc: Harrell Miscellaneous*. R package version 3.8-3. http://CRAN.R-project.org/package=Hmisc. 102

Hastie, T., Tibshirani, R., and Friedman, J. (2001), *The Elements of Statistical Learning: Data Mining, Inference, and prediction*, Springer Series in Statistics, Springer-Verlag, New York. 269, 286, 290

Hastie, T., Tibshirani, R., and Friedman, J. (2009), *The Elements of Statistical Learning: Data Mining, Inference, and Prediction*, 2 ed., Springer Series in Statistics, Springer, New York. ix, 18

Hawkins, D. (1980), *Identification of Outliers*, Chapman and Hall, London. 152

Ho, T. K. (1995), Random decision forests, in *Proceedings of the 3rd International Conference on Document Analysis and Recognition*, Montreal, Canada, IEEE Copmuter Society, pp. 278–282. 264

Ho, T. K. (1998), The random subspace method for constructing decision forests, *IEEE Transactions on Pattern Analysis and Machine Intelligence* **20**(8), 832–844. 264

Hornik, K., Buchta, C., and Zeileis, A. (2009), Open-source machine learning: R meets Weka, *Computational Statistics* **24**(2), 225–232. 286

Hothorn, T., Buehlmann, P., Kneib, T., Schmid, M., and Hofner, B. (2011), *Model-Based Boosting*. R package version 2.0-10. `http://CRAN.R-project.org/package=mboost`. 286

Hothorn, T., Hornik, K., and Zeileis, A. (2006), Unbiased recursive partitioning: A conditional inference framework, *Journal of Computational and Graphical Statistics* **15**(3), 651–674. 241, 242, 266

Inmon, W. H. (1996), *Building the Data Warehouse*, John Wiley and Sons, New York. 65

Jing, L., Ng, M. K., and Huang, J. Z. (2007), An entropy weighting k-means algorithm for subspace clustering of high-dimensional sparse data, *IEEE Transactions on Knowledge and Data Engineering* **19**(8), 1026–1041. 190

Karatzoglou, A., Smola, A., Hornik, K., and Zeileis, A. (2004), kernlab – an S4 package for kernel methods in R, *Journal of Statistical Software* **11**(9), 1–20. `http://www.jstatsoft.org/v11/i09/`. 299

Kleiber, C. and Zeileis, A. (2008), *Applied Econometrics with R*, Use R!, Springer, New York. 18

Kuhn, M., Weston, S., Coulter, N., Lenon, P., and Otles, Z. (2010), *odfWeave: Sweave Processing of Open Document Format (ODF) Files*. R package version 0.7.17. `http://CRAN.R-project.org/package=odfWeave`. 148

Kuhn, M., Wing, J., Weston, S., Williams, A., Keefer, C., and Engel-hardt, A. (2011), *caret: Classification and Regression Training*. R package version 4.76. `http://CRAN.R-project.org/package=caret`. xii

Lang, D. T. (2011), *XML: Tools for Parsing and Generating XML Within R and S-Plus*. R package version 3.4-0. `http://CRAN.R-project.org/package=XML`. 89

Lang, D. T., Swayne, D., Wickham, H., and Lawrence, M. (2011), *rggobi: Interface between R and GGobi*. R package version 2.1.17. `http://CRAN.R-project.org/package=rggobi`. 31, 137, 141

Lawrence, M. and Temple Lang, D. (2010), RGtk2: A graphical user interface toolkit for R, *Journal of Statistical Software* **37**(8), 1–52. `http://www.jstatsoft.org/v37/i08/`. xii

Liaw, A. and Wiener, M. (2002), Classification and regression by ran-domforest, *R News* **2**(3), 18–22. `http://CRAN.R-project.org/doc/Rnews/`. 249

Maechler, M., Rousseeuw, P., Struyf, A., and Hubert, M. (2005), Cluster analysis basics and extensions. 190

Maindonald, J. and Braun, J. (2007), *Data Analysis and Graphics Using R*, 2nd ed., Cambridge University Press, Cambridge. `http://wwwmaths.anu.edu.au/~johnm/r-book.html`. 18

Mingers, J. (1989), An empirical comparison of selection measures for decision-tree induction, *Machine Learning* **3**(4), 319–342. 241

Muenchen, R. A. (2008), *R for SAS and SPSS Users*, Statistics and Computing, Springer, New York. 11, 18, 68

Organisation for Economic Co-operation and Development (OECD) (1980), *OECD Guidelines on the Protection of Privacy and Transbor-der Flows of Personal Data*. `http://www.oecd.org/document/18/0,3343,en_2649_34255_1815186_1_1_1_1,00.html`. 17

Python (1989), *Programming language*, open source. `http://www.python.org`. xii

Quinlan, J. R. (1986), Induction of decision trees, *Machine Learning* **1**(1), 81–106. 241

Quinlan, J. R. (1993), *C4.5: Programs for Machine Learning*, Morgan Kaufmann, San Fransisco. 241

R Development Core Team (2011), *R: A Language and Environment for Statistical Computing*, R Foundation for Statistical Computing, Vienna. http://www.R-project.org/. vii

R Foundation for Statistical Computing (2008), *R: Regulatory Compliance and Validation Issues. A Guidance Document for the Use of R in Regulated Clinical Trial Environments*. http://www.r-project.org/doc/R-FDA.pdf. 15

Ridgeway, G. (2010), *gbm: Generalized Boosted Regression Models*. R package version 1.6-3.1. http://CRAN.R-project.org/package=gbm. 286

Ripley, B. and Lapsley, M. (2010), *RODBC: ODBC Database Access*. R package version 1.3-2. http://CRAN.R-project.org/package=RODBC. 85

Sarkar, D. (2008), *Lattice: Multivariate Data Visualisation with R*, Use R!, Springer, New York. 18, 138

Schapire, R. E., Freund, Y., Bartlett, P., and Lee, W. S. (1997), Boosting the margin: A new explanation for the effectiveness of voting methods, in D. H. Fisher, ed., *Proceedings of the 14th International Conference on Machine Learning (ICML97)*, Nashville, Tennessee, Morgan Kaufmann, San Fransisco, pp. 322–330. http://citeseer.ist.psu.edu/schapire97boosting.html. 285

Shi, T. and Horvath, S. (2006), Unsupervised learning with random forest predictors, *Journal of Computational and Graphical Statistics* **15**(1), 118–138. 265

Sing, T., Sander, O., Beerenwinkel, N., and Lengauer, T. (2009), *ROCR: Visualizing the Performance of Scoring Classifiers*. R package version 1.0-4. http://CRAN.R-project.org/package=ROCR. 312

Spector, P. (2008), *Data Manipulation with R*, Use R!, Springer, New York. 18, 68

Strobl, C., Boulesteix, A.-L., Zeileis, A., and Hothorn, T. (2007), Bias in random forest variable importance measures: Illustrations, sources and a solution, *BMC Bioinformatics* **8**(25). http://www.ncbi.nlm.nih.gov/pmc/articles/PMC1796903/. 265

Taverner, T., Verzani, J., and Conde, I. (2010), *RGtk2Extras: Data frame editor and dialog making wrapper for RGtk2*. R package. 81

Therneau, T. M. and Atkinson, B. (2011), *rpart: Recursive Partitioning*. R package version 3.1-50. http://CRAN.R-project.org/package=rpart. 215

Torgo, L. (2010), *Data Mining with R: Learning with Case Studies*, CRC Data Mining and Knowledge Discovery, Chapman and Hall, Boca Ranton, FL. ix

Tufte, E. R. (1985), *The Visual Display of Quantitative Information*, Graphics Press. 108

Tukey, J. W. (1977), *Exploratory Data Analysis*, Addison-Wesley, Reading, MA. 110

Tuszynski, J. (2009), *caTools*. R package version 1.10. http://CRAN.R-project.org/package=caTools. 286

van Buuren, S. and Groothuis-Oudshoorn, K. (2011), Mice: Multivariate imputation by chained equations in R, *Journal of Statistical Software, forthcoming* . 106

Venables, W. N. and Ripley, B. D. (2002), *Modern Applied Statistics with S*, 4th ed , Statistics and Computing, Springer, New York. 18

Warnes, G. R. (2011), *gdata: R Programming Tools for Data Manipulation*. R package version 2.8.2. http://CRAN.R-project.org/package=gdata. 51

White, D. (2010), *maptree: Mapping, Pruning, and Graphing Tree Models*. R package version 1.4-6. http://CRAN.R-project.org/package=maptree. 228

Wickham, H. (2007), Reshaping data with the reshape package, *Journal of Statistical Software* **21**(12), 1–20. http://www.jstatsoft.org/v21/i12/paper. 158

Wickham, H. (2009), *ggplot2: Elegant Graphics for Data Analysis*, Use R!, Springer. xii, 18, 137

Wickham, H., Cook, D., Buja, A., and Schloerke, B. (2010), *DescribeDisplay: R Interface to DescribeDisplay (GGobi plugin).* R package version 0.2.2. `http://CRAN.R-project.org/package=DescribeDisplay`. 147

Wickham, H., Lawrence, M., Lang, D. T., and Swayne, D. F. (2008), An introduction to rggobi, *R News* **8**(2), 3–7. 148

Williams, G. J. (1987), Some experiments in decision tree induction., *Australian Computer Journal* **19**(2), 84–91. 245

Williams, G. J. (1988), Combining decision trees: Initial results from the MIL algorithm, in J. S. Gero and R. B. Stanton, eds., *Artificial Intelligence Developments and Applications: Selected Papers from the First Australian Joint Artificial Intelligence Conference, Sydney, Australia, 2–4 November, 1987*, Elsevier Science Publishers, North-Holland, pp. 273–289. 172, 245, 264

Williams, G. J. (2009), Rattle: A data mining GUI for R, *The R Journal* **1**(2), 45–55. `http://journal.r-project.org/archive/2009-2/RJournal_2009-2_Williams.pdf`. vii

Williams, G. J., Huang, J. Z., Chen, X., and Wang, Q. (2011), *siatclust: Shenzhen Institutes of Advanced Technology Clustering Suite.* R package version 1.0.0. `http://www.siat.ac.cn`. 190

Witten, I. H. and Frank, E. (2005), *Data Mining: Practical Machine Learning Tools and Techniques*, 2nd ed., Morgan Kaufmann, San Francisco. `http://www.cs.waikato.ac.nz/~ml/weka/book.html`. 83

Wuertz, D., Rmetrics core team members, Chauss, P., King, R., Gu, C., Gross, J., Scott, D., Lumley, T., Zeileis, A., and Aas, K. (2010), *fBasics: Rmetrics—Markets and Basic Statistics.* R package version 2110.79. `http://CRAN.R-project.org/package=fBasics`. 103

Index

Made in the USA
Lexington, KY
30 August 2012